Managing Planet Earth

Managing Planet Earth

PERSPECTIVES ON POPULATION, ECOLOGY, AND THE LAW

Miguel A. Santos

Bergin & Garvey Publishers

New York • Westport, Connecticut • London

Library of Congress Cataloging-in-Publication Data

Santos, Miguel A.
   Managing planet earth : perspectives on population, ecology, and
the law / Miguel A. Santos.
     p. cm.
   Includes bibliographical references.
   ISBN 0-89789-216-X (lib. bdg. : alk. paper)
    1. Environmental law. 2. Population—Law and legislation.
   3. Environmental protection. 4. Population. I. Title.
   K3570.4.S23   1990
   344'.046—dc20
   [342.446]   89-49264

Library of Congress Catalog Card Number: 89-49264
ISBN: 0-89789-216-X

First published in 1990

Bergin & Garvey, One Madison Avenue, New York, NY 10010
An imprint of Greenwood Publishing Group, Inc.

Printed in the United States of America

∞

The paper used in this book complies with the
Permanent Paper Standard issued by the National
Information Standards Organization (Z39.48-1984).

10 9 8 7 6 5 4 3 2 1

Contents

Illustrations

Preface

Human population is presently increasing exponentially. In 1950 there were 2.5 billion people in the world. By 1970, this increased to 3.7 billion, and today there are approximately 5 billion people residing on this planet. If the present growth rate persists, there will be 30 billion people on Earth by the end of the twenty-first century. It is a truism that the human population cannot continue this infinite increased on a finite earth.

Society cannot increase beyond its ability to acquire natural resources or to dispose safely of pollutants. We must remember that interactions with other species and other humans may determine the world carrying capacity. When, how, and what laws or principles are employed by the international legal system to reach an optimum sustainable society is a matter of grave concern.

One need not be an ecologist, environmental scientist, demographer, political scientist, economist, or lawyer to understand the dangers implicit in an uncontrolled degradation of the environment. Achieving a world population restricted to the optimum number of people that the earth can adequately sustain presents problems that are difficult to solve. Calculating how many people each individual nation should have and determining the means by which politically various populations will be controlled calls for difficult decisions. Second only to the risks of international superpower confrontations stemming from interruptions in the supplies of vital resources (e.g. petroleum, chromium), the problem of achieving optimum world population will be one of the most difficult scientific-political decisions of the future.

It is a solemn thought that scientific uncertainty is so complex and pervasive in environmental decisionmaking. Science is hardly ever sure about how pollutants affect humans, biota, or the ecosphere as a whole. The relationship between the concentration of a pollutant and its effects may be linear, exponential, or even polynomial. Also, there are the additional problems of time lag and synergistic effects of interacting factors. Consequently, scientists can provide little detail on how to monitor the earth's life support systems.

Moreover, scientific methods are not capable of making subjective legal decisions because scientific principles are descriptive, not prescriptive. They do not determine how things should be; they only determine how things are, and how they will be. Scientific knowledge is essential to clarify and quantify the resources needed by society and determine what the consequences of pollution will be. Scientific principles conveyed by scientists are considered invaluable in the shaping of environmental decisions. The legal process, however, controls the direction and rate of resource consumption and habitat contamination. This limitation is quite unfortunate since the maintenance of a self-sustainable society involves the measurement of environmental stability and the balancing of it against purely economic, legal, social, and political considerations.

The international ecological crisis, with its concomitant competition for resources among nations, especially the superpowers, is our most serious threat to international world order and ecological stability. The interaction between present-day nations is a paradox, for one cannot have environmental order, stability, and sovereign nations within a larger sovereign ecosphere. At best, the larger system will be unstable. It is with the preceding considerations that this book examines the environmental crisis.

The logical development of the subject matter is divided into five chapters. Chapter 1 examines the ecological characteristics of human population, such as exponential growth and carrying capacity. This chapter also includes a discussion of the population policies in developing and developed nations. The objective of chapter 2 is to construct an analytic framework for the interaction of society with the environment. An overview of natural resources and pollution in the context of ecological process is presented. The criteria for determining the earth's carrying capacity for humans are evaluated in the third chapter. Chapter 4 proposes a synergistic model for determining the earth's carrying capacity. The final chapter, chapter 5, considers the problems and prospects of international law and environmental protection.

This book seeks to give a concrete, easily understood, and realistic analysis of the scientific and legal dimensions of environmental stability. Although intended for ecologists, environmental scientists, demographers, political scientists, economists, and lawyers, this book is not limited to them. Other academicians with an interest in the relationship between the environment and society will find this book instructive and provocative.

Managing Planet Earth

1 Human Population and Carrying Capacity

For many years humankind has justified the exploitation of natural resources with the arguments that there are plenty of resources available, and that their use is essential for growth. In the name of growth and progress, forests have been felled, water polluted, and air contaminated.

The same general argument has been used to justify population growth, and the situation has reached a crisis stage. At the present and projected growth rates, the world population will reach 6.2 billion by the year 2000 (U.S. Bureau of Census 1985). Of these, 5 billion, or 80.7 percent, will reside in Asia, Africa, and Latin America. Hence, most of the increase in the world's population is expected to occur in developing nations.

Developing nations cannot continue reproducing as they have been for the last two centuries. The sad part of the story is that some people in these nations are not aware of the overpopulation problem, particularly at the period during which couples are considering starting a family. They may also come from families or cultures with a custom that puts high value on large families. Besides, in many developing nations, effective population control programs are nonexistent even though the demand for them already exists. Understandably, many of these same people who would benefit by the reduction of fertility resent the intrusion into their private lives. There may also be problems with certain national groups who are suspicious that this is a way of further decreasing their numbers. The point was noted by Louis Ochero, who stated:

> The advocacy of family planning as synonymous with or contributing to population control, considered by the advocates as a basic requirement for economic development, was almost universally rejected in black Africa. In some areas these were viewed as thinly veiled efforts at genocide, or at best an unrealistic approach to the real issue of development.
>
> The obstacle to the spread of family planning practices in Africa is not unconnected with their foreign origin. Most of the personnel propagating family planning in Africa are either expatriate or foreign-financed. This has generated a suspicion that family planning is foreign-supported for political and even racial reasons. (Ochero 1981)

The preceding not withstanding, there is considerable data that the vast majority of parents in developing countries will practice birth control (when it is made available). At the end of this chapter we will analyze the population policies of some nations.

It should be noted that the population explosion in developing nations is not the world's only problem. Perhaps equally important is the emission of transboundary pollutants by developed nations. Pollutants such as chlorofluorocarbons (CFCs), radiation, and pesticides are all physical environmental factors that can be found in air, land, or water. They recognize no political boundaries. The activities of one nation can create pollution that is detrimental to other nations, or to all humankind. During the past few decades we have become increasingly aware that emission of some transboundary pollutants may be inadvertently and irreversibly influencing the global environment. The most recent concerns are over the greenhouse effect, acid rain, and the ozone hole in the stratosphere. (For fuller treatment of these concerns, the reader is referred to chapter 2.) As was noted by Zimmerman (1984), the Chairman of the Multilateral Conference on the Environment which was held in Munich, "Environmental pollution does not stop at frontiers and therefore we must act internationally to preserve our environment in all its diversity and beauty. . . . Next to the strengthening of peace, environmental protection is the most important task of our age."

Moreover, as we shall see in this chapter, the developed countries are responsible for the overconsumption of nonrenewable resources. It has been estimated that 6 percent of the world's population lives in developed nations. The people, however, consume 60 percent of the earth's resources (Barney 1979). Thus, we would need to multiply the population of developed nations by as many times to compare it to the population of developing nations in terms of effect on the ecosphere. As we shall see in this chapter, carrying capacity is the optimum population that could be indefinitely sustained without environmental damage. By this definition, developed nations like the United States are already overpopulated. This point was made by hearings before the U.S. Congress on the effects of population growth on natural resources and the environment:

The United States, with its bulging farm surpluses and its bulging waist-line, does not seem to be confronted with a food shortage. The world's minerals, fuels, and timber may also support an expanded population, though only at the risk of raiding the rest of the world.

But population growth presents a far broader threat than that of material resources. Too many people—particularly too many affluent people—cause air and water pollution, make noise, emit harmful chemicals, crowd open space, cause traffic congestion, and otherwise reduce the quality of life in our predominantly urban society. (Committee on Government Operations 1969)

Apparently, most of the serious transboundary pollutants and overconsumption of resources result from the activities of developed nations. Consequently, the solution to these serious problems is likely found within their borders. As the late professor of international law Wolfgang Friedmann so adequately expressed it:

In the years ahead, the active intervention of government in the economic and social affairs of nations is certain to increase greatly. This is a matter of necessity rather than ideology. The most industrialized, urbanized and congested nations will be in need of the strongest and most ubiquitous forms of governmental control. The rate of change, and the corresponding transformation of the general political ideology, is likely to be most dramatic in countries which, like the United States, have grown and prospered under the ideology of private enterprise, but now face the problems of industrialization, urbanization and pollution of environment on a larger scale than any other country. (Friedmann 1972)

Therefore, there are two kinds of international population problems. The first kind is due to the mushroom growth of population and occurs principally in the developing nations. The second kind is due to increased resource consumption with its concomitant transboundary pollutants; this kind is most obvious in the developed nations. A dilemma is that as developing nations pursue the goal of becoming industrialized, they too will contribute to the depletion of resources and pollution. This point was expressed in a report by Worldwatch Institute (1978): "As the effort to improve the welfare of a growing population places ever greater stresses on resource supplies and on the stability of ecosystems . . . pressure will have to be brought to bear on those whose activities and habits account for incongruously high proportions of . . . stresses."

It should be intuitively obvious that maintenance of the earth's resources and control of transboundary pollutants requires international laws. In a sense, much of the rest of this book is an examination of the carrying capacity of the earth and the legal administration of the ecosphere. Appropriate chapters address such fundamental questions as: Do nature's laws ultimately apply to our society? How fast is the earth's human population growing? Under what circumstances might one expect human population to grow rapidly, and then crash abruptly? What should be the government's role in population and pollution control? What variables interact to form a stable human system and ecosphere? What is the carrying capacity of the earth? What are some of the possible legal ramifications of a stable human system? How does world order relate to systematic environmental management? Is it practical or necessary to have a World Environmental Authority? With these key questions in mind, we turn our attention to the dynamics of human population growth.

The populations of different nations display different patterns of population dynamics. They may differ in birth rates, death rates, fluctuations in population size, migration, and density. By analyzing these patterns, we can predict or make logical guesses as to how fast the population will increase, and whether or not it can continue to do so ad infinitum. In this chapter we shall examine these patterns and the population policies of developed and developing nations. In order to begin our study, it is first necessary to consider some of the general properties of population ecology, notably biotic potential and carrying capacity.

BIOTIC POTENTIAL THEORY

The number of individuals of a species inhabiting a specific area (population density) rarely indicates how many there could be if the environment were favorable. This is because a population has a reproductive potential that is exponential. For example, female house mice are dependent upon their parents for 3 to 4 weeks after birth but become sexually mature at 7 weeks of age. If given the opportunity, each female may produce 11 litters in 10 months, totaling 100 or more offspring. Therefore, in four years a pair of male and female mice plus the offspring of their young would produce a total of over ten million mice. A few hundred years later the mouse population would weigh much more than the earth (figure 1-1).

According to Odum (1983), "When the environment is unlimited (space, food, or other organisms not exerting a limiting effect), the specific growth rate (the population growth per individual) becomes constant and maximum for the existing microclimatic conditions." Chapman (1928) coined the term "biotic potential" to

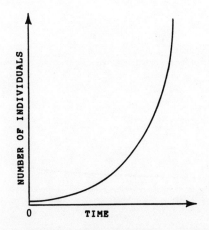

Figure 1-1. A fundamental principle of population ecology is that a population will grow exponentially when it has access to indefinite supplies of space, food, water, and other necessary resources. The slope of the curve is slight at the beginning, and then increases with time. (Adapted from Brewer 1988)

designate maximum reproductive power. Due to the biotic potential, the population of the species could double at an accelerating rate as demonstrated in figure 1-1.

In humans, the same observable fact is encountered. The human population increased relatively slowly until about 1650 A.D., when approximately half a billion people inhabited the earth (figure 1-2). The population doubled to one billion within the next two centuries, the second billion was added within eighty years, the third within thirty, the fourth within fifteen, and, in 1989, the fifth billion within twelve years. The sixth and seventh will come even faster. The increasingly shorter time indicates that humans are also capable of exponential growth.

The doubling time of a human population may be calculated by dividing seventy by the growth rate. Seventy years ordinarily is used as a demographic constant because it represents a doubling time for 1 percent growth rate. Thus, if the present world growth rate of 1.6 percent a year continues, the earth's human population will double in 43.8 years (70/1.6 = 43.8). This means that in 43.8 years, the

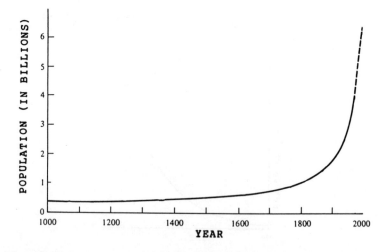

Figure 1-2. Growth curve for human population, past (solid line) and projected (broken line). In 1950, the world population stood at 2.56 billion, but by 1989 (39 years later) it had doubled to 5 billion. Projecting to the year 2000, the Population Reference Bureau estimates a world population of 6.2 billion. (Population Reference Bureau 1985)

earth will need double the amount of resources, jobs, space, etc., if the standard of living is to remain the same. Table 1-1 indicates the expected relationship between the growth rate and the doubling time by region and level of development from 1990 to 1995.

While human population growth is the result of many complex variables, the amount of population change of a nation (ignoring

Table 1-1
Projected World Average Annual Rates of Growth and Doubling Time, by Region and Level of Development: 1990-1995

REGION/LEVEL OF DEVELOPMENT	GROWTH RATE %	DOUBLING TIME (YEARS)
World	1.6	44
Developed	.5	140
Developing	1.9	37
World (Excluding Mainland China)	1.8	39
Developing (Excluding Mainland China)	2.2	32
Sub-Saharan Africa	3.0	23
Near East and North Africa	2.6	27
Asia	1.5	47
Developed	.4	175
Developing	1.6	44
Asia (Excluding Mainland China)	1.8	39
Developing (Excluding Mainland China)	1.9	37
Latin America	2.1	33
Northern America	.8	88
Europe and the Soviet Union	.4	175
Oceania	1.1	64

(Source: U. S. Bureau of Census 1985)

migration) can often be expressed by the equation:

$$C = (B - D)N$$

where C is population change, B is the birth rate, D is the death rate, and N is population size.

To gain a better quantitative idea of what this means, consider the present human population of Ethiopia. This population is now slightly over 48 million, and the birth rate is approximately 46 children born for every 1,000 persons every year, or 0.046 per person per year. The death rate is 23 persons out of every 1,000 persons per year, or 0.023 per person per year. The number of people added to every 1,000 persons each year is 46 − 23 = 23, or 0.023 per person per year. The population is therefore increasing at approximately the following rate:

$$
\begin{aligned}
C &= (0.046 - 0.023) \ (48,000,000) \\
&= (0.023) \ (48,000,000) \\
&= 1,104,000 \text{ per year}
\end{aligned}
$$

At this amount of population change, Ethiopia would acquire a population of 96,000,000 by the year 2017.

Return now to table 1-1, and notice that the doubling time differs. In each case, however, the reproductive potential of the region would eventually cause it to double in size; only the time span to achieve any given population size differs. The lesson here is simple but of vital importance. As long as the birth rate exceeds the death rate, no matter how small the difference, the population will double. This growth, plotted on a graph, produces an exponential curve (figure 1-1). However, since doubling time is a function of the growth rate, doubling time will vary greatly between nations. Moreover, it is unlikely that the population of most presently developed nations will ever double under the current stage of demographic dynamics. This is because, as explained later in this chapter, birth rates and death rates both reflect social factors, and both show a relationship to age.

CARRYING CAPACITY THEORY

At the present rate of population growth, one can determine that the sheer weight of all living humans in about two thousand years will equal the weight of the earth. Four thousand years later, it would weigh as much as the visible universe and would be "expanding outward at the speed of light" (Wilson and Bossert 1971). This biotic potential is purely a biological goal, no different from that of any other species of plant or animal on this planet. Humankind, however, is the only species that can substantially alter its environment and has shown such a growh pace.

Apparently, exponential growth cannot occur indefinitely. In nature, fortunately, large population growth of organisms (flora and fauna) are prevented by five factors: resource availability, consumers (predators, parasites, etc.), competition, self-regulation, and catastrophic events. Collectively, the factors opposing unlimited growth are known as environmental resistance.

The carrying capacity theory states that as the population approaches the level of optimum sustainable size, or carrying capacity, environmental resistance becomes greater and greater. The carrying capacity is determined by the available resources and other limiting factors in a given area. The difference between the biotic potential and the actual population growth may be a measure of the environmental resistance (figure 1-3).

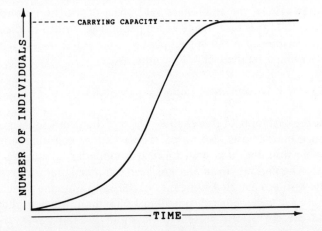

Figure 1-3. The environment is incapable of supporting a maximal rate of reproduction for any organism over an extended period of time. Environmental resistance markedly increases the death rate as well as restrains the rate of reproduction. Thus, as the population of any species approaches the level of optimum sustainable size, or carrying capacity, the environmental resistance becomes greater and greater.

The growth curve of population subject to environmental resistance differs from the exponential curve. The complete curve pattern is described by a simple equation:

$$C = (B - D)N \times (K - N)/K$$

In this equation, $(K - N)/K$ specifies that the population growth will be reduced as population number, N, approaches the carrying capacity, K. As N approaches K, $(K - N)/K$ becomes 0 and the population stops increasing. According to the equation, $(K - N)/K$ varies linearly between 0 and 1 as N increases. It assumes that each additional being exerts its environmental impact at birth and that all produce equal effects.

The logistic equation contains some important simplifications that are not characteristic of human populations. The most critical one is that each additional person makes things slightly worse for the others because of competition for available resources. The fact, however, is that this is not necessarily the case. For example, as already noted in the introduction to this chapter, each person in a developed nation consumes more resources and is therefore responsible for a greater amount of pollution. According to Ehrlich and Holdren (1971), the environmental impact depends on number of persons (N) times resource (R) used per person divided by the number of persons times the impact per unit of resource used (T), which is determined by the technology employed:

$$\text{Environmental Impact} = (N \times R)/(N \times T)$$

At the beginning of this chapter, the environmental impact of developed nations was also noted. These nations consume more renewable resources and emit more transboundary pollutants than developing nations. In fact, it has been estimated that for everyone in the world to live as Americans do, the earth's human population would have to be reduced to 0.59 to 1.13 billion, which is about 12 to 13 percent of the present population (see table 3-1). In this sense, the human population has already exceeded its sustained carrying capacity.

Another critical assumption of the logistic equation is that all individuals in a population contribute equally to the growth rate. In reality, birth rates and death rates of an entire population are affected by the relative percentages of old and young individuals and by the ratio of males to females. For example, figure 1-12 shows the age structure of developing nations. Notice these nations have extremely high percentages of younger age classes. Thus, in the case of the developing nations, because of a high percentage of young individuals, the world's human population would continue to increase well into the next century, even if all families of reproductive-age individuals had only two children from now on.

The logistic equation contains a third important assumption that is not a characteristic of human population. It assumes that each person exerts its environmental impact at birth. Because people grow during their lives, their impacts on others and on the environment increase. For example, newborns may have little effect on the ability of other people in the population to consume resource until the babies reach a size where they consume a large amount of resources.

It is possible to recognize two general types of growth curves: the S-shaped and J-shaped curves (Odum 1989). The S-shaped growth curve is produced when the environmental resistance becomes increasingly effective as the density of the population rises. For the sake of analysis, four phases of the S-shaped curve are distinguished: positive acceleration, negative acceleration, stationary, and deceleration (figure 1-4).

The positive acceleration phase is characterized by the population growing exponentially. During this phase, $N = 0$ or is very small; hence, $(K - N)/K = 0$ or is very small. Before long, the environmental resistance causes the positive acceleration phase to change to a negative acceleration phase. During this phase, N is much greater than 0 but much less than K. The environmental resistance becomes increasingly effective as the density of the population rises. That is, the environmental resistance becomes density dependent, or the intensity of the resistance is partly determined by the density of the population.

The rate of increase slows down as the population grows older and approaches the optimum population of individuals that the habitat and/or nutritional resources can support. When the rate of reproduction equals the rate of death, the overall number of individuals remains constant. This is referred to as the stationary phase. During phase $N = K$; therefore, the expression $(K - N)/K = 0$. At this time, the population is stable and has achieved

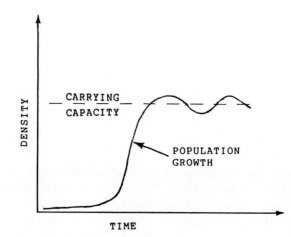

Figure 1-4. Idealized S-shaped population growth curve. The population tends to stabilize at a level of equilibrium that may be thought of as the carrying capacity.

zero population growth (ZPG). In some species, such as fruit fly and paramecium, the population levels off at the carrying capacity. In others, such as bacterium and yeast, the exhaustion of nutrients in the environment and/or accumulation of toxic wastes may cause a curve deceleration of the population growth. During this phase, N is less than K, and the death rate exceeds the rate of reproduction.

The J-shaped curve is characteristic of the population growth of small insects with short life cycles and of annual plant populations. In this type of curve, the population increases in density at a rapidly accelerating rate until the environmental resistance suddenly comes into play, at which point the population density falls dramatically (figure 1-5). For the sake of analysis, two phases of this curve are distinguished: positive acceleration and deceleration.

During the positive acceleration phase, the population grows exponentially until it strikes the line set by the habitat and/or nutritional resources. The deceleration phase is a density-independent effect of the environmental resistance. During deceleration the population declines abruptly because it has exhausted the available resources.

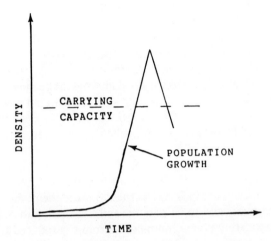

Figure 1-5. In the J-shaped curve, the population erupts and overshoots the carrying capacity, exhausting a temporarily abundant resource; this is followed by a sudden precipitous population decline. In this type of curve, there is a long delay before forces operating to control the rate of population are fully effective.

Every species shows a tendency toward a rise in number. Some follow an S-shaped curve, while others follow a J-shaped curve, and still others a mixture. Even though the environment has neither the habitat resource nor the nutritional resource for all possible off-spring of any species, the inherent readiness to overreproduce is a characteristic of life. The only variable is the time it takes to reach a large population size. Since there are many causes of death in nature, including predators, parasites, floods, starvation, and diseases, excess reproduction is valuable to a species since many of the offspring fail to survive to maturity.

Like other species, humans have encountered environmental resistance. But, unlike other species, humans have the ingenuity to avoid or postpone the effects of environmental resistance. As a result, human population has been spurred by three "revolutions," each conquering some form of environmental resistance (Deevey 1960).

The culture revolution may have been the first advance that allowed our population to enter a phase of exponential growth.

This included the discovery of fire and the development of tools, weapons, shelter, and protective clothing. With this, primitive people could more effectively modify and exploit their environment. The use of weapons and fire opened up a new range of foods. Cooking, for example, can kill food pathogens, remove volatile plant toxins, and can also soften plants and make them more digestible. With protective clothing and shelter, mankind was capable of penetrating and establishing itself on all of the continents. Ten thousand years ago, the earth probably supported about five million humans.

Cultivation of plants and animal husbandry may have allowed a second surge of growth. Starting around 8000 B.C., people domesticated plants and animals and learned to breed them; they learned to till, to fertilize, and to irrigate the soil. According to Sears (1957), by domesticating plants, humans enormously reduced the land required for sustaining each person by a factor on the order of at least 500. As a result of the much more dependable sources of food that became available, human population began to grow rapidly, climbing to an estimated 130 million people by the time of Christ.

The third population growth surge began in Europe between the sixteenth and seventeenth centuries. The factors responsible for the continuing increase into the twentieth century are not fully understood for all regions of the world. For many regions, however, it is highly probable that three major factors were involved: industry, agriculture, and medicine. The Renaissance in Europe during the seventeenth century, with its renewed interest in science and medicine, ultimately led to the establishment of industry and the development of medicine, methods of sanitation, and other means of disease control. Agriculture was being improved and crop failures were less frequent. The discovery of antibiotics and the development of many new vaccines in the twentieth century further reduced a major limiting factor on human population growth. This population explosion continues today, and in many respects, is the source of our current international environmental crisis.

WORLD POPULATION GROWTH

A rough approximation of the history of human population is seen in figure 1-2. In 8000 B.C., the world was inhabited by some 5-10 million humans. Slow and steady increases continued until 1

A.D. when the population stood at 100-160 million. By the late eighteenth century, the number had grown slowly and unsteadily, increasing to less than one billion. With the arrival of the industrial revolution, the trend changed to a more stable increase. Currently, the earth's population is more than five billion. The present population obviously represents an exponential stage, and from our earlier discussion of growth curves we can speculate that it is either a J-shaped curve or the first part of an S-shaped curve (positive acceleration phase). Although it is impossible to gauge whether the worldwide growth curve is a J- or S-shaped curve, the population growth is very uneven from one nation to the next.

Population Growth in Developed Nations

The growth curves of Europe, North America, and the U.S.S.R. are shown in figure 1-6, from the 1800s to the present. These developed regions increased their population size primarily because of a decline in the death rate. This decline is attributed to technological development that has, in turn, increased food production, reduced

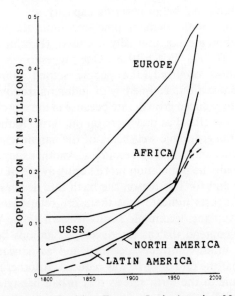

Figure 1-6. Population of Africa, Europe, Latin America, North America, and U.S.S.R., 1800 to 1987. (Source: U.S. Bureau of Census 1987)

the effects of the physical environment, made parasites less abundant than before, and improved medical care. Although these nations exhibit some self-regulatory mechanisms, like warfare, this has had little significant effect on their population trends.

Compared to developing nations, the developed nations have experienced a reduction in growth rates. This reduction is due principally to a decline in fertility. The people in these regions are opting for smaller families. In the 1940s, demographers noticed that a decline in the developed nations' birth rate eventually follows a reduction in its death rate. This decline in the birth rate is known as the demographic transition (Notestein et al. 1963, Coale 1974).

In the demographic transition, a nation's growth rate goes through three phases. In the first phase, the growth rate is low because of a high death rate (especially among infants) which cancels out the effects of a high birth rate (figure 1-7). In the second phase, as the nation begins to develop, the growth rate is high because of a lower death rate. The result is that the population begins to grow rapidly. In the third phase, the growth rate is again low because the birth rate declines. The population enters a steady state or equilibrium, meaning that the population may oscillate somewhat above and below carrying capacity.

The reasons for the demographic transition are poorly understood, and they are not mutually exclusive (Hardin 1968, Coale 1970, Winikoff 1978, Frisch 1978). One suggestion made is that as industrialization occurred, the people settled more in urban centers. Urbanization, coupled with industrialization, may have contributed to reduced growth rate because in the city children were no longer as beneficial as they were on the farm. Rather than contributing to the economic well-being of the parent, children represented a serious drain on their resources. Another suggestion is that city living made the population more keenly aware of the concerns of crowding and for this reason the birth rate decreased. In addition, some reports indicate that there are relationships between family size and socioeconomic status (Brown 1976, World Bank 1984). They contend that as the developed nations became more affluent, as infant mortality declined and as educational levels increased, that the rate of birth was reduced. Another reason given for the decline in the birth rate is that urban lifestyle, combined with industrialization, provides more career alternatives for

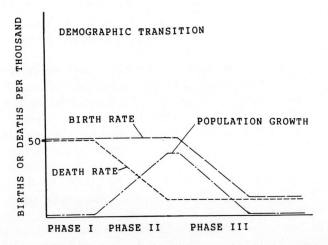

Figure 1-7. According to the demographic transition theory, populations have three phases depending on their developmental stage. Developed nations are nearing the end of the third phase while most developing nations are still in the early stages of the third phase. (Adapted from Nebel 1987 and Clapham, Jr. 1981)

women. The relationship between marriage, family size, and working in women's lives was described by Westhoff (1978):

> Imagine, ultimately, a society in which men and women have the same incomes, in which there are as many women as men who are lawyers, engineers, corporation executives, physicians, and salespeople. What would the consequences be for marriage and fertility? . . . The decline of fertility, then, can be regarded as both a cause and a consequence of changes in marriage and the family. Having fewer children or none can be construed as facilitating the economic activity of women and increasing their economic equality with men, two developments that in turn make marriage and childbearing less of an automatic social response. A future increasingly incompatible with the concept of traditional marriage is almost certain to be a future of low fertility in the developed countries.

Regardless of the reasons for the decline in the growth rate, the fact remains that in most developed regions, populations have more or less stabilized. Currently, some developed nations, such as

Austria, Belgium, Denmark, East Germany, Finland, France, Italy, Luxembourg, Norway, Sweden, the United Kingdom, and West Germany are at, or near, zero population growth (U.N. 1988).

Europe and the Soviet Union are presently the home of approximately 770 million people, or 15.9 percent of the world population. It is projected that by the year 2000, Europe and the Soviet Union will have nearly 815 million people, or approximately 13-14 percent of the world population. Northern America (Canada and the United States) will also increase in population from 264 million today to 297 million in the year 2000. It will also decrease in terms of percentage; presently, Northern America makes up 5.5 percent of the world population, but by the year 2000 it will make up only 4.8 percent.

The United States has one of the highest growth rates of all industrialized nations. Its population is increasing by about 2-3 million each year. Total growth is projected to be 32-36 million for the remainder of the 20th century. According to the U.S. Bureau of Census projection, the population will reach 250-260 million by the year 2000. As figure 1-8 indicates, the population appears to be growing linearly. There are two important reasons why the U.S. population is not leveling off. The first can be associated with the "baby boom" of 1947-1964; during this time parents had larger families than today. This resulted in the "baby boom" and the mo-

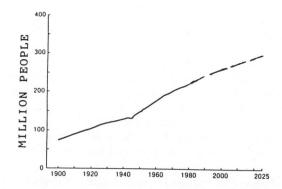

Figure 1-8. Total U.S. population, 1900 to 1987, and projected to 2025 (broken line). Projections based on a birth rate of 2.1 children per woman. (U.S. Bureau of Census 1987)

mentum in population increase that has not yet subsided. For example, in 1980 the women who had reached their childbearing years had increased by more than a million as compared to 1970. Even though these women were averaging 1.9 children each, because there were more of them, the total number of births grew from 3.1 million per annum in the mid-1970s, to 3.6 million in 1980-1981. The second factor influencing the population increase in the U.S. is immigration; the immigration rate is not easily ascertained. Registered or legal immigrants number about 500,000 per annum, but rough estimates of unregistered or illegal immigrants range from 100,000 to as high as one million per year. (In a "normal" year, about 125,000-150,000 persons leave the U.S.) Both legal and illegal immigrants contribute approximately 40-50 percent of the total annual population growth in the United States.

The growth rate of the U.S. population, however, has been declining since the 1950s (figure 1-9); from 1980 to 1985 it was about 1 percent. If current trends continue, zero population growth (ZPG)— the point at which the population simply replaces itself—could be approached by the year 2030. Similarly to other developed nations, the principal reason for this decrease is the number of births. On the average, women now have only 1.8 children during their childbearing years. In 1940, the average was 2.5. In the late 1950s it was 3.6. The replacement level fertility, or how many infants are born to each woman during her reproductive years that would result in

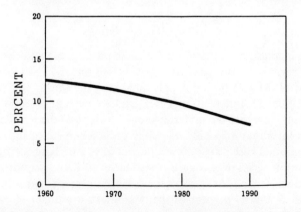

Figure 1-9. U.S. population growth rates, 1960 to 1987. (U.S. Bureau of Census 1987)

replacement (but no growth) of the reproducing population, is 2.1 children. Replacement level fertility is slightly greater than 2 because parents must replace themselves and replace offspring who die before reaching sexual maturity.

A growing U.S. population will tend to increase the pressures on natural resources, such as clean water, air, and land. The reduction in growth rates, coupled with increasing longevity, is causing the average age of the population to rise (figure 1-10). What are the implications of an older population? To quote from the Report of the Commission on Population Growth and the American Future (1972):

One concern often expressed about an older age structure is that there will be a larger proportion of the population who are less adaptable to political and social change, thus suggesting the possibility of "social stagnation." Others have suggested that Sweden and England, both of which have older structures than ours, are not especially slow to change; but, it is difficult to generalize from particular cases. In any event, other factors, such as accumulated wealth and level of education, obscure the relationship between chronological age and resistance to change [Coale 1972]. For example, older generations typically grew up in an era of less education; this gap will narrow in the future. . . . In summary, we are led to the conclusion that the age structure of population is unlikely to be decisive in the forms of social organization which emerge. And, as we have seen, there are many advantages of population stabilization which seem clearly to outweigh any fears of an older population.

Population Growth in Developing Nations

The increase in human population is much greater in the developing regions of the world, as figures 1-6 and 1-11 demonstrate. Today, of the 4.85 billion people inhabiting the world, 58.5 percent live in Asia, 11.4 percent in Africa, and 8.4 percent in Latin America. Thus, together the underdeveloped regions of the world are the homes for about 3.79 billion people, or 78.3 percent of the world's population. These developing countries have a high momentum for population growth because they have a high proportion of individuals who have not reached childbearing age. In these nations, about 40 percent of the inhabitants are under 15 years of age, whereas in developed nations, the corresponding proportion is

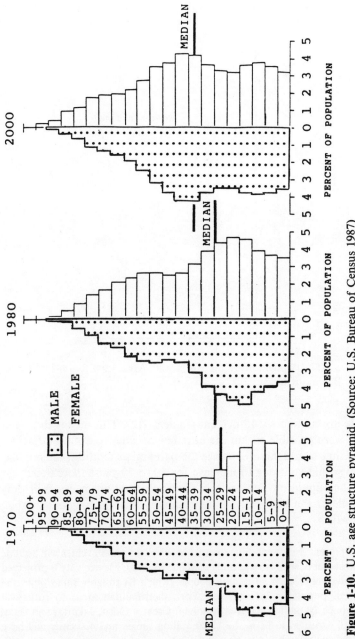

Figure 1-10. U.S. age structure pyramid. (Source: U.S. Bureau of Census 1987)

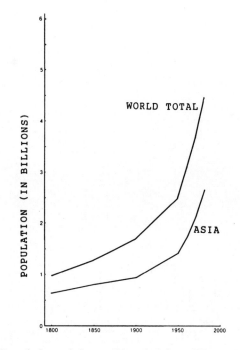

Figure 1-11. Population of the world and Asia, 1800 to 1987. (Source: U.S. Bureau of Census 1987)

only approximately 15 percent (figure 1-12). This means that even if it were possible within the next few decades to obtain a fertility rate that would merely replace the parental generation, the population would continue to increase for 50 to 70 years thereafter.

The population explosion in developing nations is not their only problem. Perhaps equally important is their uncontrolled urban growth:

> If present trends continue, many [developing nation's] cities will become almost inconceivably large and crowded. By 2000, Mexico City is projected to have more than 30 million people [which is] roughly three times the present population of the New York metropolitan area. Calcutta will approach 20 million. Greater Bombay, Greater Cairo, Jakarta, and Seoul are all expected to be in the 15-20 million range, and 400 cities will have passed the million mark. (Barney 1979)

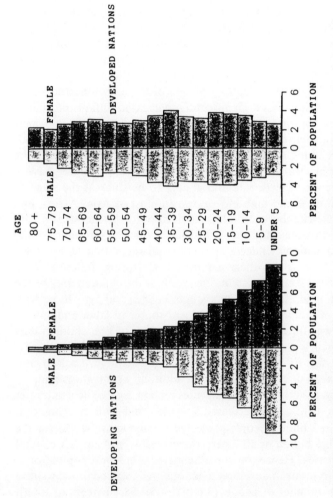

Figure 1-12. Comparison of age structure pyramids between developing and developed nations. A broad-based pyramid characterizes most populations in developing nations of Africa, Asia, and Latin America. A bell-shaped pyramid characterizes most populations in developed nations of Europe, North America, and the Soviet Union. (Source: U.S. Bureau of Census 1987)

Most of these cities have grown rapidly and created a new kind of relationship with the surrounding areas. One of the immediate impacts of urbanization is the competition for agricultural land and the pressures it exerts on flora and fauna. Moreover, urbanization will put extreme pressures on sanitation, water supplies, health care, food, shelter, and jobs.

Growth rates are highest in much of Asia, Africa, and Latin America due to high degrees of fertility and dramatic decreases in mortality which immediately followed World War II. Improved living conditions such as potable water, better nutrition, and better medical services have significantly decreased infant mortality, allowing many more children to reach sexual maturity and have families of their own. Unlike the developed regions of the world, most developing nations are not proceeding smoothly through the demographic transition. The developing regions did not have a major decline in birth rate and appear to be stuck at the stage of low death rates and high birth rates. In most of Asia, Africa, and Latin America, the people have not enjoyed the increase in economic well-being that the industrial revolution brought to the developed nations. Indeed, the industrial revolution has scarcely penetrated some of these countries. Offspring function as a type of social security system, inasmuch as the children may be the only economic assistance for parents in their old age. In agricultural regions, children are a significant source of labor and, consequently, wealth. Lacking the very necessities of life, children may be the parent's only wealth and source of pride. Culturally, there is often prestige for the parent who has many children. Lack of education and birth control devices, in addition, aggravate the impact of extreme poverty. Thus, birth rates remain high while death rates have been significantly decreased by science and medical technology.

Of the 6.2 billion people projected for the entire world for the year 2000, 5 billion or 80.7 percent will reside in Asia, Africa, and Latin America. Hence, most of the increase in world population is expected to occur in developing nations. Asia will continue to have the largest number of people (3.6 billion or 58.1 percent of total), followed by Africa (850 million or 13.7 percent), and Latin America (557 million or 9 percent). Growth rate in Africa has been stable (and high) while Latin America and Asia both show declining trends.

The resources of developing nations, which grow more slowly, have to be divided among a population that is growing very rapidly, with the result that living conditions are declining. According to Lester R. Brown, today we face a new problem:

Not only have [developing nations] failed to complete the demographic transition, but the deteriorating relationship between people and ecological support systems is lowering living standards in many of these countries, making it difficult for them to do so.

The risk in some countries is that death rates will begin to rise in response to declining living standards, pushing countries back into the first stage. In 1963, Frank Notestein pointed out that "such a rise in mortality would demonstrate the bankruptcy of all our [development] efforts." For a number of countries, that specter of bankruptcy is growing uncomfortably close. (Brown 1987)

WORLD POPULATION POLICIES AND PROGRAMS

Homo sapiens is the only species that has shown an ever-increasing rate of population growth. Other species do not show this tremendous population increase because they are regulated by the environment. We have noted that, with the development of modern medicine, death rates in most countries dropped dramatically, but continuing high birth rates along with the declining death rates have caused the population size to mushroom in many poor countries.

It is obvious that some population policy is necessary. Unlike wild animals that do not have the intelligence to foresee signs of environmental resistance, for mankind the alternative is a voluntary self-regulation of our species. The growth attitude may have served its purpose to compensate for the many humans that once died during earthquakes, pestilence, famines, and wars. Today, however, these factors have been reduced and no longer take a large toll. We must begin to voluntarily reduce our own population by increasing mortality or reducing natality. The former method was practiced by some ancient peoples, such as the Spartans, who left many infants to die. The latter system, however, is more humane and more easily accomplished since we currently have several birth control methods, such as the intra-uterine device (IUD), birth control pills, voluntary sterilization, abortion, and so forth.

The government involvement in population policy is not entirely theoretical. In fact, voluntary and involuntary methods are currently being used, or being seriously considered, by many countries. International conferences, for example, in Bucharest, Romania in 1974, provide an opportunity to discuss issues of population and development and to prepare a population plan of action. At the Bucharest conference, the developed nations encouraged the developing nations to formally initiate programs to help minimize the effect that overpopulation can place on economic development. At that time very few developing nations had population policies or programs. Besides, many of these same nations who would profit by reduction of the birth rate resented intrusion in their jurisdiction and suspected that the developed nations, who were trying to influence them, were acting for selfish reasons. Many third world countries contended that only a "New International Economic Order" would solve their development problems (CEQ 1984). In spite of their different points of views, the countries at Bucharest adopted the World Population Plan of Action (WPPA), which has served as a comprehensive model for national and international organizations in formulating and implementing population policies and programs.

Ten years later, the United Nations International Conference of Population convened in Mexico City in August 1984. The purpose of the conference was: (1) to evaluate progress made in the implementation of the WPPA; (2) to establish priority actions and objectives to expedite its implementation; and (3) to strengthen and sustain the momentum already caused by population activities (CEQ 1984).

Due to the comprehensive nature of the conference and to the intrusion of extraneous political issues, the conference achieved limited goals. One of the chief political debates occurred when the United States argued that population growth and the free market system play a role in increasing economic development. At the 1984 conference the United States, in stating its official policy, said:

Population growth is, of itself, neither good nor bad. It becomes an asset or a problem in conjunction with other factors, such as economic policy and social constraints. While many nations face critical problems associated with population pressures, we do not face a global population crisis. The United States reaffirms its long-standing commitment to economic and voluntary family planning assistance programs for developing nations. At the same time, it will take advantage of the experiences of the past two decades to measure the effectiveness of its economic assistance while ensuring that its family planning funds are used in ways consistent with human dignity and familial values.

Population and economic development policies are interrelated and mutually reinforcing. Based on historical experience, the twin objectives of economic growth and population stability without compulsion will most readily be achieved through adoption of market-oriented economic policies that encourage private investment and initiative. Such policies result in the most rapid increases in standard of living, which in turn result in a lowering of birth rates as parents opt for smaller families. (CEQ 1984)

Furthermore, the United States expressed some concerns regarding coercion to achieve population control. The United States issued a policy statement that reaffirmed its long-standing commitment to voluntary family planning assistance programs for developing countries; however, the statement noted that:

The United States supports the primacy of the right of couples to determine the number of children they will have, and it does not accept abortion as an acceptable element of family planning programs. Accordingly, the United States will tighten existing restrictions on the use of its population assistance funds. It will no longer contribute to any nongovernmental organization that performs or actively promotes abortion, nor will it provide family planning funds to any nation that engages in coercion to achieve its population objectives. (CEQ 1984)

The delegates affirmed (perhaps an overstatement) the United States position and displayed broad agreement. The delegates agreed that there is an interrelationship between population and economic development in improving the living conditions of the world's inhabitants. All regions reaffirmed the major tenets of the WPPA. Moreover, it was restated that both individuals and couples have the right to make informed decisions about childbearing and that it was important to "respect cultural and religious values" (CEQ 1984).

As a general rule, the countries in which population policies have been most effective have four factors in common: (1) an effective national population education program; (2) widely available family planning services; (3) a combination of incentives for small families or disincentives for large ones; and (4) widespread improvements in economic and social conditions (Jacobson 1987). These four factors are developed in the remainder of this chapter.

Very few families understand the concept of carrying capacity and ecosphere stability. The national government can alleviate the difficult decision-making process of families by maintaining population control clinics to provide access to contraception, sterilization, abortion, and other birth control technology. Some proponents have even argued that the government should "repeal restraints to homosexual unions between consenting adults."

In a voluntary national population education program, the decision to procreate is left to each individual couple. The government may play a role by discouraging couples from reproducing when the nation is overpopulated, and encourging reproduction when it is underpopulated. (All developed nations except the United States have a system of family allowances which provide payments to parents for support of their children.) Nevertheless, the final decision is made freely by the man and woman who are considering having children. The population control program would be informative and advisory, but not coercive. Such programs respect cultural and religious values and are in full agreement with the United Nations Universal Declaration of Human Rights (1967), which states: "The Universal Declaration of Human Rights describes the family as the natural and fundamental unit of society. It follows that any choice and decision with regard to the size of the family must irrevocably rest with the family itself, and cannot be made by anyone else."

Voluntary birth control programs should be the primary method for achieving a stable population. Because some nations are already too overpopulated, other means of reducing growth have been undertaken in some countries. A study of the population policies of India and China is useful in illustrating both the successes and difficulties in implementing family planning programs.

India was the first nation to establish a national family planning program in 1951-52, although it was not pursued seriously until 1965-66. In the 1970s, India was the first government to offer cash incentives or prizes (including portable radios) to men who had fathered several children if they would accept sterilization. Parts of India have employed deferred incentives in the form of old-age pensions and medical care, and economic incentives to be paid in the future to couples who have succeeded in maintaining a small family. India's family-planning program was claimed to be a major factor in reducing birth and fertility rates by about 50 percent. Nevertheless, due to the backlash against the compulsory programs, the government began to relax its stringent policy. In fact, the strict policy was one of the key factors leading to the overthrow of Indira Gandhi's government in 1977 (Mauldin 1980). Beginning in 1978, largely in response to the changing public attitudes, a new approach was initiated. The Indian government raised the age at which marriage is allowed from 18 to 21 years for males and from 15 to 18 years for females. In the 1980s, it has been noted that the population growth rate is again increasing, and the Indian government is again considering renewing its population control policy.

China, the largest developing nation, with over 1 billion people (one fifth of the world's population), currently has the strongest national policy on population control of any nation. According to Teitelbaum (1975), the legal minimum age for marriage was raised to 20 for women and 22 for men. In addition, the government provides free birth control measures, financial, educational, housing, and retirement inducements for couples that have only one child. In some regions of China, forced sterilization is a component of the program. The population-control policy also imposes penalties for having more than two children: those who refuse to abort a third pregnancy must pay a fine. Between 1970 and 1986, China's population growth rate dropped from 2.6 percent per year to about 1.1 percent per year, comparable to that of the United States. The objective of the Chinese government is zero population growth by the year 2000 (Mauldin 1980). Whether such efforts will succeed remains to be seen; there is concern that as the population of China ages it may be necessary to encourage births to supply needed workers.

In conclusion, considerable data on changes in fertility in relation to changes in average income and adult literacy have been collected. The data show that in nearly every country where average income and literacy have increased, fertility has dropped (World Bank 1984). Recall from our earlier discussion of the demographic transition that economic and social improvements may have the effect of limiting population growth. This point is described by W. Mauldin:

Social setting or development has a substantial relationship to fertility decline, certainly on a holistic basis and probably selectively with regard to health and educational status. Family-planning programs have a significant, independent effect, certainly in developing countries with favorable social settings and also under conditions in countries with less favorable settings (including the three largest: China, India, and Indonesia). Moreover, the longer the program and the clearer its demographic intent, the greater its effect. Social setting and family planning programs together predict or "explain" a large part of fertility decline. (Mauldin 1980).

Consequently, birth-control programs alone, without education and economic opportunity for the majority of a country's population, have rarely been effective. As the President of the World Bank, Barber Conable (1988) wisely suggested, developing countries must "renew and expand efforts to limit population growth," by raising the living conditions. Apparently, many developing countries still suffer from extreme poverty, which Conable describes as a "moral outrage." Despite the need for reducing fertility, about 27 percent of the developing nations have not yet established family planning programs.

The ideal birth-control programs are those that permit maximum freedom and diversity (Berelson 1969). That is, those programs that consider goals other than population control, such as improved health care, and those that do not weigh heavily on the disadvantaged. It is erroneous to assume that people in developing countries would not recognize the relationship between the overpopulation problem and their economic welfare. A recognition of the problem should be encouraged. If this is done, the people of these nations themselves would recognize the problem. For religious and other legitimate reasons, mentally competent parents may refuse to practice birth control; the state should not force them to do so. The integrity of the ecosphere may be in danger of deteriorating, yet, we must remember that the patient in population control programs is not the individual, but rather the ecosphere. A greater, even more present danger is the intrusion of the state into the privacy of all our personal lives.

I cannot help but to ponder on the thoughts of Justice Brandeis' dissenting opinion in *Olmstead v. United States* (1928):

Experience should teach us to be most on guard to protect liberty when the Government's purposes are beneficient. Men born to freedom are naturally alert to repel invasion of their liberty by evil-minded rulers. The greatest dangers to liberty lurk in insidious encroachment by men of zeal, well-meaning but without understanding.

2   *Natural Resources, Pollution, and Carrying Capacity*

In recent years the realization that the human population is depleting the planet's natural resources and polluting the environment has generated a strong interest in the study of human environments. The study of humans and their environment covers a very broad range of topics. The human environment, for example, includes the combination of physical and biological conditions that affect and influence mankind, as well as the complexity of social and cultural conditions that affect the nature of an individual or the community. Thus, in order to study humans and their environment one would need to know art, biology, chemistry, ecology, economics, engineering, physics, psychology, sociology, or just about all of the disciplines found in a university curriculum.

Therefore, it is suggested for academic convenience, that we disregard the distinction between humans and their environment and instead consider humans and their environment as an interacting unit called the "anthroposystem." This system is a functional and structural unit of interwoven and overlapping hierarchies of organization, which maintains human civilization in space and time. An anthroposystem is a structural and functional unit of the ecosphere because it can be considered a self-contained system, provided it has an energy source.

What few people seem to realize is that more progress has been made in dealing with discrete factors of our environment (e.g., with air and water pollution) than with the relationship between humans and their environment. This, of course, is due primarily to the fact that everyone can see and feel the immediate effects of pollutants. However, it tends to obscure the most important and fundamental issue facing the ecosphere. There has been a failure in the identification, measurement, and inclusion of all the variables interacting to form the various ecosystems and human systems.

It is a gloomy thought that our environment may be so complex in space and time that a universal concept such as the anthroposystem must be either imaginary or unrealistic. To avoid the pitfalls of treating life-support systems as oversimplified black boxes, scientists must develop universal theories that clarify, quantify, and unify sustainable human environment systems. That is, we must develop testable theories that explain and predict the behavior of the "forest" (the entire human-environment system), not just the trees.

The objective of this chapter is to construct a system framework for the interaction of humankind with the environment. The discussion focuses on some unifying principles directly related to understanding a sustainable human system. An overview of natural resources and pollution in the context of the international environmental crisis is presented. This chapter also reviews the concept of stability as it applies to the human society. The specific details of how to determine carrying capacity and manage the international global commons are explored in later chapters.

CONSTRUCTING A SYSTEM FRAMEWORK OF THE ENVIRONMENTAL PROBLEM

Humankind belongs to a species of primate called Homo sapiens. Humans appeared on earth somewhere between 300,000 and 100,000 years ago (Nelson and Jurmain 1985). Their ancestors were probably ape-like primates that lived in Africa. They developed artifacts, social order, and communication systems which enabled them to cope with environmental factors. Like many animals, humans were nomads in the beginning, constantly shifting their habitations in order to find food and water. Several races of mankind are still nomadic, such as some bushmen of Africa and the Australian aborigines.

It is not difficult to imagine that some group of men encountered areas where there was plenty of food and water. Such areas most likely gave rise to villages, since there is evidence that Homo sapiens had created villages 9,000 years before Christ. The aggregation of many people in relatively small areas probably attracted many scavenger animals, such as dogs and pigs. Dogs were the first animals to be domesticated by humans in approximately 12,000 B.C. (Harlan 1976). Sheep, goat, cattle, pig,. donkey, buffalo, camel, and llama were domesticated shortly thereafter.

Once settled, the villagers hunted for local animals and gathered local plants for food. As they gathered the plants, it is likely that some seeds fell near their primitive shelters. Some villagers probably observed that seeds gave rise to plants, and soon they began to gather seeds in order to cultivate them. In fact, there is evidence that agricultural societies appeared by about 7500 B.C. These societies cultivated wheat and barley. Soon after, other seeds were cultivated, such as rice in Asia and corn in the Americas. Thanks to the increased production of food, made possible by domesticating animals and plants, large numbers of people were able to live together in permanent settlements for the first time. As a consequence, this permitted the development of towns and cities.

One of the first cities to appear was Catal Huyuk, in Turkey, around 6500 B.C. (Harlan 1976). Early cities created regional activities such as farming, manufacturing, financing, and governing. The city and its associated local communities made possible the development of a self-sustainable human system, or an anthroposystem.

We can divide the anthroposystem into components on the basis of how they influence mankind's chance of survival in a stable environment. When this approach is followed, the anthroposystem may be broken down into four major components which then can be separately defined and logically analyzed. Figure 2-1 illustrates the four components of an anthroposystem. They are matrix, producers, consumers, and decomposers. The matrix in an anthroposystem is composed of all the non-living and non-productive parts of the system such as buildings, streets, air, and water. The matrix provides the edifice or fabric in which the other components operate. The component which manufactures or yields products is termed producer. The producers include two categories: agri-

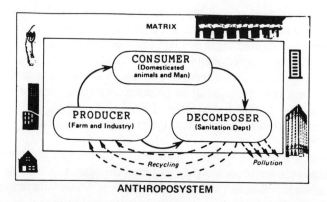

Figure 2-1. The anthroposystem is one of the simplest constituents of the ecosphere which operates to form a functional stable unit. The diagram demonstrates the relationship of the four components of the anthroposystem with the arrows pointing to some of the major pathways of materials.

cultural and industrial. The agricultural producers are green plants, such as wheat, barley, rice, and corn. The industrial producers are the machines and tools operated by humans to produce shelter, clothing, transportation, and so forth. The consumers consist of humans and their domesticated animals. The decomposers in the anthroposystem are the waste-water treatment plants, resource recovery plants, electrostatic precipitators, spray collectors or scrubbers, etc. A fully functional decomposer component would serve to maximize the recovery of resource. Today, however, the resources that are of no need are dumped into surrounding environments rather than recycled. This dumping causes contamination of the environment and the depletion of natural resources.

Since different human environments or societies grade almost imperceptibly into one another, a division into anthroposystem is a convenient way in which to organize our thinking. There is, however, a level of human organization that is larger than the anthroposystem, yet is easier to see as a united whole. This is the nation, state, or country. However, air, water, pollutants, and other molecules and compounds do not stop at national boundaries. An anthroposystem exists in the ecosphere and as such should be viewed in an ecological context. The boundary between an anthroposystem and its surroundings is an imaginary one used only for convenience of discussion, such as the imaginary boundary around a mole of glucose molecules in a solution, a transportation system, or a health care system. These are abstract systems without a fixed boundary. An anthroposystem is an artificial system produced by human efforts and exists as a result of these efforts.

The term ecosystem refers to the interacting system of organisms and their nonliving habitat. An ecosystem was originally defined by Tansley (1935) as the "whole system (in the sense of physics) including not only the organisms complex, but also the whole complex of physical factors forming what we call the environment." Recently, Odum (1983) gave a more comprehensive definition of an ecosystem as "Any unit (a biosystem) that includes all the organisms that function together (the biotic community) in a given area interacting with the physical environment so that a flow of energy leads to clearly defined biotic structures and cycling of materials between living and nonliving parts in an ecological system or ecosystem." The sum total of life on earth, together with the global environment and the earth's total resources, make up what LaMont C. Cole (1958) called the ecosphere. Thus, we can argue that there are two fundamentally different units in the ecosphere: human systems and ecological systems (Santos 1974).

Ecosystems (ponds, forests, etc.) and anthroposystems grade almost imperceptibly into one another, but a division into ecosystems and anthroposystems is a convenient way of organizing our thinking as long as we do not regard them as actually being distinct or internally homogeneous. For obvious reasons, the divisions are mentally constructed and designed in order to cope with the tremendous diversity of the environment. They are not laws of nature; the environment does not come into two convenient categories labeled ecosystem and anthroposystem. For example, the nitrogen element which forms a part of your eyes may have been a part of a plankton's membrane that was consumed by the fish you ate. Thus, there is an interrelationship of parts between and within ecosystems and anthroposystems. The real world is composed of a mosaic of interrelated natural ecosystems and artificial anthroposystems.

In an interrelated or coupled system, different variables affect one another, like cogs in a machine; turning one part causes motion everywhere else. The Council on Environmental Quality (1985), for instance, explained eloquently the complexity of the environment:

As our understanding of the physics, chemistry, and biology of the Earth has progressed, more and more evidence has accrued that the separate components of the planet are really not separate at all, but are interrelated in both obvious and complex ways. Ultimately, we cannot separate the geosphere (oceans, atmosphere, ice cover, and solid earth) from the biosphere (aquatic and terrestrial, including man), because events in one part of the Earth system are intimately related to happenings elsewhere. As a simple example of this linkage, consider rocks and rain reacting to form soils, while plants and animals through their mutual activities, in turn, influence the composition of both the soil and the atmosphere.

Viewing the world from a very different political perspective, but reaching a similar conclusion, was Eduard Shevardnadze (1988) foreign minister of the Soviet Union. He stated in an address to the U.N. General Assembly that "the dividing lines of the bipolar world are receding. The biosphere recognizes no division into blocs, alliances, or systems. All share the same climatic system."

NATURAL RESOURCES AND CARRYING CAPACITY

An anthroposystem is a structural and functional unit of the ecosphere because it can be considered a self-contained system as long as it has an external energy source. It is hypothetically a self-sufficient unit, and its components cannot survive apart from the whole. That is, the four components of an anthroposystem (producer, consumer, decomposer, and matrix) merely represent subsystems of a sustainable human system.

Once an anthroposystem exceeds to the point where its demand begins to exceed the sustainable yield of resources and the assimilative capacity of its matrix for pollutants, it has exceeded its carrying capacity. This in turn reduces production, triggering a change in the system's structure and function. Consequently, the concept of carrying capacity focuses on interactions between society, its activities, and the surrounding matrix. It highlights natural thresholds that might otherwise remain unclear. (As we shall see in chapter 3, the criteria for determining the carrying capacity can be classified into two basic types: socioeconomic and ecological.)

An ideal anthroposystem satisfies its needs without diminishing the prospects of surviving in space and time. Judging by this measure, contemporary society fails to meet this criterion. The decomposers in human systems are not as developed as in natural ecosystems. Ecosystems rely on their decomposers to break down dead plants and animals and to recover recycled waste materials. Human systems recover very few of their waste materials (domestic, industrial, agricultural); they are simply dumped into the environment where they accumulate and may cause pollution. The parasitic and destructive nature of human systems results in the depletion of natural resources and the pollution of the environment. Leaving aside the issue of pollution until next section, let us consider the problem of resource depletion.

A natural resource is that which is depended upon for aid or support of the anthroposystem. These resources are essential for the survival of the anthroposystem and also for determining the size and organization of society. However, as indicated by the sociologist William R. Catton, Jr. (1987), "With different organizations and technologies, one population of humans can be a very different sort of ecological entity than another human aggregate."

There are two categories of natural resources: functional and habitat. Functional resources include factors required for the biological needs of humans and their domesticated organisms and for industrial processes. They provide energy and/or the materials required for the metabolism of the system. Functional resources include agricultural nutrients, solar energy, limestone, chromium, aluminum, iron, oxygen, water, and so forth. Habitat resources include factors not required for energy and/or materials, such as temperature and space. Some resources can serve both habitat as well as functional resources. Solar energy, for example, warms the earth's surface and drives the hydrologic cycle and the winds. Moreover, a small percentage powers photosynthesis.

Resources have been described as renewable or nonrenewable resources. Renewable resources can either be made usable by resource recovery or can be replenished by natural processes. Some examples of renewable resources are food crops and animals, wildlife, forests, and other living things, as well as fresh air, fresh water, and fertile soil. Nonrenewable resources cannot be made new or be restored to a former sound condition. Mineral resources (such as iron and copper) and fossil fuels (gas, petroleum, and coal) are nonrenewable resources. Each removal from its origin is final and irrevocable. These resources exist in a fixed supply or cannot be replaced as fast as they are used.

If the rate of consumption or loss of a given renewable resource exceeds the maximum replenishment or harvesting for long, the stocks will be depleted. The human system dependent on the stocks will be impoverished and may perish. In much of the world, for example, there are often droughts because consumption of water exceeds rainfall. Moreover, as Freese (1985) recognized, when resources needed by a consumer population are replaced at an approximately constant rate, replacement rate is exceeded by consumption rate. Resource depletion steadily influences further availability, so that relative scarcity increases logarithmically, and as time proceeds, environmental damage becomes less and less reversible. It is argued that today's societies have consistently misinterpreted rates of discovery with rates of replacement (Pratt 1952, Simon and Kahn 1984). As explained by Catton, Jr. (1987): "By definition, the replacement rate for nonrenewable resources must be effectively constant, i.e., zero, and any nonzero rate of use must exceed it." The lesson here is simple but of crucial importance. As long as the consumption rate in a population exceeds the replacement rate, no matter how small the difference, the relative scarcity will grow exponentially.

When material resources are recovered, the human system is a closed system for that particular resource. This is because no significant exchange of resource occurs between that system and its surroundings. There is a complete treatment of all wastes so that there is no discharge of pollutants into the environment. In such a closed system, the resource can theoretically last indefinitely.

The real world differs significantly from a closed ideal anthroposystem model. First, in every manufacturing process, raw materials are extracted, refined, processed, and transformed into finished products. Along the way, these processes generate large quantities of wastes. Such wastes are inevitable by-products of industry. Second, because of the second law of thermodynamics, energy use is never completely efficient. This natural law, a conclusion based on observation of the natural world, says that whenever energy is converted from one form to another, some potential energy is lost (see chapter 4). For example, a car converts the chemical energy stored in gasoline to the kinetic energy of movement. In the process, 75 percent of the energy is immediately lost as heat. The same thing occurs in organisms. As glucose is used for energy by organisms, only 40 percent of the energy is captured as biologically useful energy; the other 60 percent of the chemical energy from glucose is dissipated as heat. A consequence of the second law of thermodynamics is that energy cannot be recycled or reused; therefore, real anthroposystems are open systems, as far as energy is concerned, in theory and in fact. Thirdly, as mentioned previously, the ecosphere is composed of a mosaic of interrelated human and natural systems. Thus, human systems tend to be open systems since resources are both imported and exported (figure 2-2).

The importing and exporting, as well as the internal transfer rate of a resource, can be described mathematically. An anthropo-

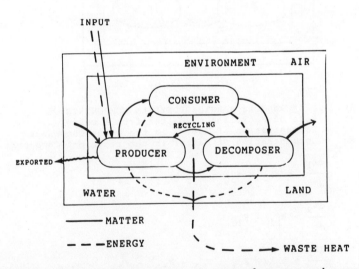

Figure 2-2. Real human systems are open systems for matter and energy. An anthroposystem can be visualized as a synthesis of sustainable human systems that can be represented as a complex of feedforward cascades of energy flow and matter recovery between the four components (producer, consumer, decomposer, and matrix). Input refers to the amount of energy or matter introduced into the system for storage, conversion of kind, or conversion of characteristics.

system can be defined by a set of coupled differential equations, one for each of the four components, expressed in terms of matter cycling and energy flow between components. The equations for matter cycling are provided in figure 2-3.

A root cause of the international environmental problem is that certain resources, such as water and air, do not recognize political boundaries and freely migrate throughout the whole ecosphere.

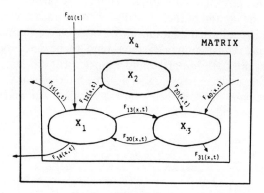

$$X_1 = F(X_1, X_2, X_3, X_4, t)$$

X_1, X_2, X_3, and X_4 = ANTHROPOSYSTEM COMPONENTS
F_{ij} = Rate of Recycling of Matter from i to j
X_i = Concentration of Matter in the Donor Component
X_j = Concentration of Matter in the recipient Component

$$\frac{dX_1}{dT} = F_{01} + F_{30} - F_{12} - F_{13} - F_{14} \ F_{15}$$

$$\frac{dX_2}{dt} = F_{12} - F_{20}$$

$$\frac{dX_3}{dt} = F_{20} + F_{13} + F_{40} - F_{30} - F_{31}$$

$$\frac{dX_4}{dt} = F_{15} - F_{40}$$

Figure 2-3. Cycling of matter through an anthroposystem (not all pathways are included).

Moreover, these transboundary resources are "free" goods with a price of zero. That is, they are resources that do not belong to any special individual or nation but belong freely to the whole international community.

The biologist Garret Hardin (1968), in his well-known article, was one of the first individuals to point out that a major cause of resource depletion is the assumption that air and water are free goods. According to him, where there is common ownership of a resource, no person assumes responsibility for husbanding it. Hardin used the example of cattle grazing on a public pasture:

Picture a pasture open to all. It is to be expected that each herdsman will try to keep as many cattle as possible on the commons. . . . As a rational being, each herdsman seeks to maximize gain. Explicitly or implicitly, more or less consciously, he asks; "What is the utility to me of adding one more animal to my herd?" This utility has one negative and one positive component.

1) The positive component is a function of the increment of one animal. Since the herdsman receives all the proceeds from the sale of the additional animal, the positive utility is nearly + 1.

2) The negative component is a function of the additional overgrazing created by one more animal. Since, however, the effects of overgrazing are shared by all the herdsmen, the negative utility for any particular decision-making herdsman is only a fraction of − 1.

Adding together the component partial utilities, the rational herdsman concludes that the only sensible course for him to pursue is to add another animal to his herd. And another; and another. . . . But this is the conclusion reached by each and every rational herdsman sharing a commons. Therein is the tragedy.

Obviously, if every nation in the world comes to this same conclusion, there is no hope for the global commons. When a nation consumes resources without considering the global damage, it disperses this negative attitude throughout the whole international community. Yet when it consumes the resources and manufactures a product, this positive component is a gain for itself alone. Individual nations are encouraged to minimize their costs irrespective of the ecospheric damage which may result. When the principle of economics operates, the damage that is done to the resources is not calculated into the cost-benefit analysis. They are, therefore, external or outside the direct cost of manufacturing; economists call them externalities.

POLLUTION AND CARRYING CAPACITY

Pollution, or habitat contamination, occurs because the anthroposystem does not have a mature, well-developed decomposer component. Much of what we have come to call pollution is in reality the nonrecoverable matter resources (figure 2-3, F_{31}) and waste heat (figure 2-2). Waste heat, or thermal pollution, is used to designate a man-induced alteration of natural water temperature. Steam electric power plants are the most common source of thermal pollution. These plants discharge large volumes of warm water that have been used as a cooling agent in the process of generating electricity. This type of pollution can have disastrous consequences on aquatic organisms (Langford 1972). Increased temperature, in general, decreases the level of dissolved oxygen available to aquatic organisms. Heated water may also increase toxic effects of certain pollutants. Furthermore, most organisms are able to exist only within a certain temperature range. Therefore, increased temperatures would tend to kill many aquatic species.

Pollution is one of the most pressing environmental crises. By introducing pollutants such as pesticides, radioactive isotopes, and heavy metals into our air, land, and water, the growth of industries and cities has placed tremendous burdens on the environment. There are now approximately 65,000 different chemicals in the marketplace, and new ones are being invented at an ever-increasing rate. In addition, the use of already precious supplies of energy and materials for such luxuries as motor boats, air conditioners, hair dryers, etc., has added more stress to the environment by depleting these limited resources. Any use of natural resources at a rate higher than nature's capacity to restore itself can result in pollution of air, water, and land.

From the standpoint of the international environmental crisis, it is convenient to recognize four general characteristics of pollutants:

1. Pollutants recognize no political boundaries.
2. Many toxins cannot be degraded by organisms and consequently persist in the ecosphere for many years.

3. Pollutants destroy biota and habitat.

4. Development of an international policy to control pollutants is hindered by the uncertainties associated with their effects.

Transboundary Pollutants

Pollutants such as chlorofluorocarbons (CFCs), radiation, pesticides, carbon dioxides and other gases are all physical environmental factors that can be found in air, land, or water. They are capable of moving from one sphere to another. Very often, chemicals in the land are carried by rain water to nearby waterways where they may cause pollution. Pollutants know no national boundaries. The activities of one nation can create pollution that is detrimental to other nations or to all humankind.

The transboundary movements of pollutants lead to unusual and complex economic and political difficulties. Individuals in one nation may suffer economic loss and health hazards as a result of pollution originating in another nation, yet they may not benefit from the economic activity that causes the pollution. On the other hand, a nation or jurisdiction that acts to minimize pollution which is being transported over hundreds of miles may gain little local environmental benefit. The interjurisdictional problems related to the regulation of long-range air pollution are especially apparent in the acid rain issue, which has led to several international disagreements (for example, between the United States and Canada).

As already noted, the transboundary issue has been politically divisive due to socioeconomic costs and benefits which accrue to different nations. The root of this problem is the "Tragedy of the Commons" (Hardin 1968). Consequently, one of the most important adjustments to be made in the operation of the international economy is the internalization of the ecospheric cost when calculating the ratio of cost to benefit. (The significance of economic externalities was mentioned in the previous section of this chapter.) This fact underscores the need for international cooperation in the management of the global commons. Nevertheless, no comprehensive international legal agreements controlling transboundary pollutants exist.

In 1979, members of the U.N. Economic Commission of Europe ratified a Convention of Transboundary Air Pollution. The thirty-four nations, including the United States, pledged to limit, and as far as possible, gradually reduce, and prevent air pollution. Because this ratification is a movement in "the right direction" (Schware and Kellogg 1983), it constitutes a preliminary step toward an international implementation. For example, as Rosencranz (1980) observes, "The agreement does not compel abatement action. It includes no mechanism for enforcement of its terms, nor does it delineate the responsibility of member states to abate pollution causing damage in another state or to award compensation for such damage."

In addition to international agreements, there are a few cases which deal with transboundary pollutants. In the classic case of the Trail Smelter Arbitration (1941), the tribunal was asked to deal with a smelter in British Columbia and the effect of its fumes on a number of farms in the State of Washington. The tribunal concluded:

[U]nder the principles of international law, as well as of the law of the United States, no State has the right to use or permit the use of its territory in such a manner as to cause injury by properties or persons therein, when the case is of serious consequence and the injury is established by clear and convincing evidence. . . .

Considering the circumstances of the case, the Tribunal holds that the Dominion of Canada is responsible in international law for the conduct of the Trail Smelter. Apart from the undertakings in the Convention, it is, therefore, the duty of the Government of the Dominion of Canada to see to it that this conduct should be in conformity with the obligation of the Dominion under international law

It has been pointed out, however, that the Trail Smelter Arbitration is sue generis:

The tribunal imposed a detailed regime of controls over the emissions of SO₂ fumes from the smelter. The case however, is perhaps of questionable validity as precedent since Canada had specifically assumed international liability for damage caused to the United States from activities within Canada and because it was agreed by both countries to establish a special binational tribunal to "arbitrate" the amount of damages. (Nanda and Moore 1983)

International agreements and court cases are based on the Roman legal maxim "sic utere tuo ut alienum non laedas" or as defined "Use your own property in such a manner as not to injure that of another" (*Black's Law Dictionary* 1979). However, as Rosencranz (1983) astutely notes: "Unfortunately, neither principles nor maxims are of much consequence in the case of transboundary air pollution. Nations rarely relinquish jurisdiction over cases of pollution emanating from their territory."

We shall highlight a few of the problems of international law and the management of the global commons in chapter 5 after the basic principles governing system behavior, natural resources, and pollution have been outlined.

Apparently, as we have already stressed, the solution to transboundary pollutants is not likely found within one nation's borders. Moreover, it is becoming increasingly apparent that the transport of pollutants over long distances raises the question of global pollution such as ozone depletion, acid rain, and the greenhouse effect. The following paragraphs outline some of the international implications of these problems.

Ozone and Radiation. The ozone problem is among the Earth's most ubiquitous environmental concerns. During the past decade or so, there has been particular interest in analyzing the processes that control atmospheric ozone, since it has been predicted that human activities may inadvertently and irreversibly deplete the ozone layer. So far, it looks like ozone depletion is linked to a combination of meteorological factors and chlorofluorocarbons (CFCs). This phenomenon was first observed as an "ozone hole" in the stratosphere over Antarctica (Farman et al. 1985). In 1988, scientists reported that the global ozone layer may be declining as well (Bowman 1988). In order to understand the problem of the depleting ozone shield, the danger of ultraviolet radiation should be discussed first.

The sun emits a large amount of ultraviolet radiation, but the earth's ozone layer absorbs the radiation and thus screens life from this kind of solar radiation. This is fortunate because UV radiation is absorbed by nucleic acids (genetic information is stored in nucleic acids). The effects of UV radiation involve the excitation of molecules such as nucleic acids, which then form cross links. These distortions of nucleic acids interfere with protein synthesis and also cause mutations and cancer. Consequently, it is predicted that many life forms will be damaged by excess radiation (Maugh 1980, Heck et al. 1983, Skarby and Sellden 1984, Cohn 1987). The epidermis, the outer skin covering that covers the more sensitive inner tissue, protects organisms from too much radiation. Moreover, in some species, such as humans, specialized cells in the skin produce melanin. Melanin is the pigment responsible for skin color. Light-skinned people have cells that produce smaller quantities of melanin than do the cells of dark-skinned people. This characteristic is inherited. The production of melanin, however, can be increased by exposure to the sun, resulting in the bronze coloring of the skin known as suntan. In the process, the pigment absorbs much of the UV radiation and thereby protects the skin from further injury. However, if the skin is overexposed to the sun, the pigment cannot absorb all the UV radiation, and the skin is injured. The skin becomes inflamed and over the years UV exposure can cause the skin to get wrinkled. Moreover, the radiation can cause dividing cells to become cancerous.

As ozone concentrations decrease, the intensity of biologically damaging ultraviolet radiation in natural daylight increases. A 16 percent ozone decrease will produce an increase in UV radiation of about 44 percent at mid latitudes, and a 30 percent ozone decrease would increase radiation approximately 100 percent. The National Academy of Sciences predicts that a 16 percent ozone depletion would eventually cause several thousand more cases of melanoma per year in the United States. Many of those would be fatal. The depletion would also cause several hundred thousand more cases of other skin cancers. A 16 to 30 percent depletion of ozone is likely to reduce crop yields of several kinds of plants. Larval forms of several important seafood species, as well as microorganisms at the base of the marine food chain, would suffer appreciable killing as a result of a 16 to 30 percent ozone depletion.

Given the current state of scientific uncertainty about ozone depletion, the effect of human activities on ozone is largely unknown. Whatever the cause, it is apparent that this problem "could literally alter the course of life on earth" (Brown and Wolf 1987).

The "ozone hole" issue has been difficult to resolve because all these issues (ozone depletion, acid deposition, and greenhouse warming) have been associated to some degree with changes in atmospheric composition. For several decades, scientists have sought to understand the complex interplay among the chemical, radiative, and dynamic processes that govern how ozone is formed in the atmosphere.

Oxygen, in addition to forming the stable molecule O_2, can also exist in another exceedingly reactive molecular form called ozone (O_3). The characteristic odor that can often be detected around certain electrical equipment is caused by ozone. Even in very small concentrations, ozone can be easily recognized by its characteristic odor. In sufficient quantities, ozone may cause impairment of the function of the lungs, irritation of mucous membranes, and complications for those with cardiorespiratory diseases. Ozone is also formed when nitrogen dioxide (NO_2) absorbs ultraviolet radiation and splits into nitric oxide (NO) and one oxygen atom (O); in subsequent reaction O combines O_2 to form O_3. Ozone formed by this mechanism is currently believed to be one of the major constituents of photochemical smog. Another method that forms ozone in limited quantities in the upper atmosphere is direct absorption of ultraviolet radiation by oxygen. Oxygen splits into two oxygen atoms ($O_2 \rightarrow O + O$) that subsequently combine with O_2 to form ozone ($O + O_2 \rightarrow O_3$). Ozone formed by this natural mechanism forms a protective layer in the stratosphere, shielding the earth against the sun's harmful ultraviolet radiation.

Numerous scientists are studying the effects of the continuing release of CFCs into the atmosphere; these CFCs are causing deterioration of the earth's protective layer of ozone. CFCs are a class of chemical compounds used as solvents, aerosol propellants, blowing agents in foam, and refrigerants. CFCs deplete ozone in the atmosphere, and as a result, there is an increase in damaging UV radiation. (The chlorine released from CFCs is believed to reduce stratospheric ozone concentrations; one molecule of a CFC destroys thousands of molecules of ozone.)

The United States, Canada, Sweden, and Norway banned nonessential use of CFCs in aerosol spray cans in the 1970s. Other countries, such as Austria, Australia, Japan, New Zealand, and Switzerland, have established controls over or have voluntarily limited production of CFCs. The result of ongoing worldwide research efforts may establish a basis for international banning of CFCs.

International organizations like the United Nations Environmental Programme (UNEP) are concerned with the ozone depletion, particularly recognizing that effective control needs a coordinated global ratification. In 1982, delegates from about twenty nations met in Sweden under UNEP auspices to begin a discussion on the requirement of an international framework for the protection of the ozone shield which would:

(1) harmonize regulatory control actions on ozone modifying substances at the international level, (2) increase coordination of ozone related research, and (3) increase the exchange of information on all scientific, technical and legal issues relevant to the ozone issue. (CEQ 1985)

As of 1989, the United States and 35 other nations have signed the treaty regulating the use of CFCs. According to this treaty, by 1998, production of CFCs will be cut to half of its 1986 level. However, according to Shea (1989):

Although an impressive diplomatic achievement and an important first step, the agreement is so riddled with loopholes that its objectives will not be met. Furthermore, scientific findings subsequent to the negotiations reveal that even if the treaty's goals were met, significant further deterioration of the ozone layer would still occur.

Scientific uncertainty also complicates the decision-making process. One hypothesis suggests solar influences are the culprit in ozone depletion (Vallis and Naturajan 1987), while another attributes the phenomenon to stratospheric clouds (McElroy et al. 1986). Nevertheless, most scientists link ozone depletion to CFCs. Moreover, as already noted, the Earth's environment is a coupled system. For example, continued CFC release could eventually cause a slight warming of the earth's surface—in addition to that predicted from carbon dioxide buildup.

Ground-based sources of nitrous oxide, which can become oxides of nitrogen in the upper atmosphere, are also believed to reduce ozone concentrations. High-flying supersonic aircraft (SSTs) are a potential source of oxides of nitrogen in the sensitive part of the ozone layer (about 10 miles above the earth's surface), but the current and projected numbers of these aircraft pose little threat.

Ozone concentrations in the lower atmosphere appear to be increasing due to photochemical smog while ozone concentrations in the upper atmosphere are decreasing due to CFCs, SSTs, or other phenomena. Hence, the total amount of ozone in the air might not change significantly. It is not clear, however, whether these two processes are exactly balancing each other out. Presently, researchers are attempting to establish more accurately the effects of various chemicals on ozone levels.

Acid Rain. Acid rain (or, more accurately, acid deposition) is a complex and controversial phenomenon in which human-made and natural emissions, particularly sulfur dioxide and nitrogen oxides, are chemically transformed in the atmosphere to produce acid compounds (sulfuric and nitric acids). These compounds fall back to earth in the form of aerosols and particulates (dry deposition), or by raindrops, snowflakes, fog, or dew (wet deposition). According to Postel (1984), North America and Europe may now be receiving as much as 30 times more acidity than they would if precipitation fell through a clean atmosphere.

Recently, the role of oxidants such as ozone and hydrogen peroxide (which result from emission of nitrogen oxides and volatile organic compounds) has assumed a greater importance because they have an effect on the atmospheric processes that transform sulfur dioxide into sulfuric acid, and nitrogen oxides into nitric acid. Moreover, trace metals commonly found in fuels can act as a catalyst in the atmospheric transformation of SO_2 to sulfates (Rahn and Lowenthal 1984).

Although those atmospheric processes are not entirely understood, the series of chemical reactions result in the formation of sulfuric and nitric acids in the air (Patrick et al. 1981, Schindler et al. 1985, Galloway et al. 1987). The pH of unpolluted, natural precipitation often lies between 5.0 and 5.6. Presently, however, the pH in some regions often approaches 4.0, perhaps 10 times lower than normal. Acid rain has been associated by various scientific studies to the acidification of some lakes and streams and consequently to a reduction in fish populations. In addition, acid deposition may deplete mineral nutrients in the soil; the mobilization of toxic metals in solid and water bodies, the reduction of forest and agricultural yields, and the damage to historic buildings and monuments constructed of marble or limestone are all negative effects of acid deposition.

Once injected into the atmosphere, much of the sulfur and nitrogen is usually deposited within a short distance. However, a portion of these emissions is transported hundreds or thousands of miles by prevailing winds, and undergoes complex transformation into acidic products described previously. These products may result in significant environmental damage regardless of political boundaries. The international transport of the acidic products has caused concern in a number of different regions. For example, a large amount of the acid precipitation that is damaging the environment of northeastern United States and eastern Canada is caused by industrial activity in the American midwest. The United States is working closely with Canada to develop a mutually acceptable process of alleviating this problem. It is likely that this concern will remain high on the future agendas of the governments of these two nations.

The acid problem has also perplexed scientists and public officials in Europe (Ottar 1977). Declines in fish populations in Scandinavian rivers have been attributed to acid deposition. Germany is concerned about the possible effects of acid rain on parts of the Black Forest, since more than a third of the forest is dying. A number of Western as well as Soviet bloc countries have expressed concern about the possible effects of acid rain on lakes and streams, forests, agricultural crops, and historic monuments. Many have agreed to take steps to implement a reduction in the level of sulfur dioxide emissions.

Spurred by the dramatic effect of acid rain on the ecosphere, 21 nations led the push for sulfur dioxide control. In the mid-1980s, negotiations took place on a sulfur dioxide protocol under the Long-Range Transboundary Air Pollution Convention in a U.N. Economic Commission for Europe. The protocol generated by this convention called for a 30 percent reduction of sulfur dioxide emission based on 1980 levels starting in 1993. According to the U.S. Council on Environmental Quality (1985), the United States did not ratify the provision, "because the U.S. had achieved substantial sulfur dioxide emissions during the 1970s, in contrast to most other countries, and this reduction would not have been recognized." Unfortunately, this philosophy is a reflection of the tragedy of the commons.

Implications of the Greenhouse Effect. The possibility of global climate change induced by an increase in pollutants in the atmosphere is potentially the most important international issue facing humanity. Yet, decision-makers have a very difficult time dealing with this concern because of scientific uncertainty and because legal action concerned with the warming of the earth might well carry them too far from the fundamental problems of most of their constituents. In order to understand the political problem, the scientific uncertainty will be stressed first.

The composition of the global atmosphere is a major determinant of the earth's climate. The process of maintaining the earth's temperature within fixed ranges occurs, in part, because the sun and the earth interact to create something that can be likened to a greenhouse. Thirty percent of the incoming radiation of the sun is scattered or reflected by clouds or by the earth's surface. Twenty percent is absorbed by oxygen, ozone, water vapor and droplets, and dust. The remaining 50 percent of the incoming radiation finally reaches the ground or ocean where it is absorbed as heat. A very small percentage (less than 1 percent) of the light energy that arrives at the surface of the earth is converted to chemical energy by plants.

The absorbed heat that is finally given off by the earth does not readily escape into outer space. This heat is absorbed by carbon dioxide and other substances in the atmosphere that cause the greenhouse effect as was previously mentioned (Watson et al. 1986). That is, the sun, as a hot body, emits short wavelength radiation which is not readily absorbed by these substances. The earth, as a cooler body, emits long-wave radiation which is then absorbed by atmospheric substances. This same principle occurs on a sunny day in a car with closed windows. The shorter wavelength of light readily penetrates the windows and warms the seat and dashboard. The longer wavelengths of radiation emitted by the seat and dashboard do not readily pass through the windows. Hence the inside of the car gets hot (greenhouse effect).

Recently, humans have been adding more carbon dioxide and other gases to the atmosphere and have been heating up the environment by their various activities at work, at home, and at play. Another disturbing worldwide trend is the continuing loss of forests to urban and other developments which also affect global climate. The combined effect of carbon dioxide, heat pollution, and deforestation is the increased warming of the earth's atmosphere.

Every human activity involves energy. The energy we need to live and work is suplied to us through the food we eat—dietary energy. We need other fuels too, like fossil fuels; fuel energy is used to reduce physical labor and to provide light, heat, transportation, and so forth. As dietary and fuel energy are burned, much of the energy is dissipated or diluted as waste heat. In highly populated urban centers, this waste heat is enough to cause inner city temperatures to be 5-20°C warmer than adjacent rural areas. On a worldwide scale, however, this waste heat is less significant. It has been estimated that in order to raise the global temperature 1°C, humankind would have to increase its energy consumption by 100-fold.

An increase in carbon dioxide and other gases, however, is expected to have a significant effect on the global temperature. This is because carbon dioxide and other gases, such as nitrous oxide, ammonia, sulfur dioxide, etc., slow down the rate of heat loss from the earth by absorbing infrared radiation. The combustion of fossil fuels is one of the primary causes of increased atmospheric concentrations of carbon dioxide and other gases (Trabalka 1985). Woodwell (1978a), however, estimates that it is the burning of forests and changes in the organic levels in the soils that are subject to deforestation and cultivation, and these are currently more important than fossil fuels. Substantial scientific evidence, however, indicates that the prime cause of the release of carbon dioxide into the atmosphere is probably the burning of fossil fuels.

It has been estimated that atmospheric carbon dioxide levels have increased by 15 to 25 percent since 1800. As shown in figure 2-4, carbon dioxide concentration has increased linearly since 1958. Projections of future increases in fossil fuel combustion indicate that the amount of carbon dioxide could double over the next hundred years ending in the middle of the next century.

Two kinds of effects of deforestation are noteworthy. First, changes in the earth's climate might result from alterations in the global albedo, that is, the reflection of light and heat from the earth's surface, when light-absorbing forests are removed. (The percentage of the total radiation of a planet that is reflected from its surface is called albedo. The average albedo of the earth is 30 percent.) The heat balance of the earth would change, producing consequent changes in wind and rainfall patterns. Second, forests

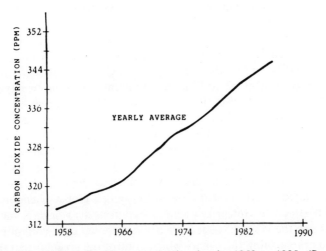

Figure 2-4. Carbon dioxide concentration in air, 1958 to 1985. (Source: Bacastow and Keeling 1981, CEQ 1985)

are a great storehouse of carbon; roughly half of the carbon in the earth's biomass is stored in forests. With large net losses of forests, the concentration of carbon dioxide in the atmosphere could rise by 30 percent, adding to the already-rising trend present, as the result of burning fossil fuels.

While the warming is continuing, man's activities are also interfering with the environment in another way—by emitting particles into the atmosphere (Mitchell, Jr. 1978). It is feared that this increase of particulate matter in the atmosphere may scatter or reflect back into space about 30 percent more of the sun's radiation than is currently being reflected, resulting in a cooler climate.

If the particulates are making the earth cooler but the increased carbon dioxide, other gases, and heat are making it warmer, what direction will the earth's temperature ultimately take? The cooling of the earth could speed up the inevitable ice age. On the other hand a rise of 2° to 3°F in a period of three to four decades could lead to both a thermal expansion of oceans and the melting of the polar ice caps (Henderson-Sellers and McGuffie 1986). As a result, sea level would rise and flooding may occur in many coastal and low-lying areas. The sea level might be raised by 400 feet, and dust bowls and desert temperatures would be created over most of the world.

According to the data on temperature, the average temperature has increased 1 °F over the past one hundred years. A number of recent studies suggest the possibility of a global warming of 1.5 to 4.5 degrees Celsius (3 to 8 degrees Fahrenheit) in the twenty-first century (NAS 1979). Environmental effects associated with climatic change would not be wholly adverse. While some parts, such as the midwestern United States, the Soviet Union, and China would become drier, other regions would benefit from a global warming period since it would most likely enhance agriculture, forestry, and water availability (Manabe and Wetherald 1986).

As stated above, greenhouse warming is a global phenomenon which affects the integrity of every nation in the world. Some nations like the United States have actively shown their commitment to comprehending the events which control the greenhouse effect, by funding research that should minimize the scientific uncertainties which presently exist concerning the magnitude of projected temperature change for different global regions. However, meaningful effort to minimize global climate modification requires international laws. The governments of Canada, the Netherlands, Norway, the U.S., and the Soviet Union have expressed interest in international negotiatons (Brown et al. 1989). But, as with so many major environmental concerns facing policymakers, the timing and fortitude of their response is significant:

> Unfortunately, the challenge of making global warming a central concern of national energy planners is far from being realized. Energy policymaking is often driven by self-interested industries and unions, and some, such as the oil and coal lobbies, push for policies that accelerate global warming. Key legislative committees are dominated by provinces or states that produce fossil fuels; many of the laws and tax breaks that emerge are intended to propel their growth. In Eastern Europe, ossified energy ministries continue to emphasize meeting their five-year plans regardless of the ecological costs.
>
> The tendency is simply to add global warming to a long list of considerations that go into making energy policy. This is not enough. If energy policymaking continues to be the domain of short-term thinking and narrow political considerations, there can be little hope. If the climate is to be stabilized, it must become the cornerstone of national energy policies. (Brown et al. 1989)

Decomposition of Pollutants

The processes responsible for the alteration or disintegration of pollutants may, for convenience, be labelled biological and physico-chemical. Biological disintegration, decomposition of pollutants through the biological activities of organisms, has long been cited as a cause of disintegration of pollutants. There are essentially two classes of pollutants in relation to their decomposition by living organisms: biodegradable and nonbiodegradable. Biodegradable wastes are capable of being broken down or used by organisms. They include organic waste products, phosphates, and inorganic salts. Nonbiodegradable pollutants resist decomposition by organisms, and remain in the environment for indefinite periods of time. Nonbiodegradable pollutants include bottles, cans, metal, plastics, certain pesticides and herbicides, and radioactive isotopes.

Biodegradable pollutants are temporary nuisances which organisms break down into harmless compounds. If the pollutant is organic (carbohydrates, proteins, fats, nucleic acids, etc.), the organism obtains energy and/or material for its own use in the process of breaking it down. However, biodegradable pollutants could have serious environmental results if a large quantity is released in a small area. For example, dumping of organic or food waste in a small pond depletes the oxygen supply. The fish have no available oxygen left; hence, they die. Thus, biodegradable substances become pollutants when they are dumped into the environment because they cannot be broken down by organisms at such a fast pace. Ecologists use the term "assimilative capacity" to express the ability of an aquatic ecosystem to assimilate a substance without degrading or damaging its integrity (Cairns 1978). Integrity is generally defined as maintenance of structure and function characteristics of the system.

Nonbiodegradable pollutants are dangerous partly because the organisms have neither evolved enzymes capable of digesting them, nor have they developed a defensive system against them. The fat-soluble nonbiodegradable pollutants, such as methylated mercury, chlorinated hydrocarbons (DDT, PCB, etc.), benzene, and poly-aromatic hydrocarbons, have an additional, and more significant, property. Since they are fat soluble, but not water soluble, these pollutants are not excreted in urine but are instead accumulated in the fat of organisms. Since they cannot metabolize these toxins, the organisms retain 100 percent of these pollutants. Toxins can, therefore, be concentrated within organisms far beyond their environmental levels. The process is called biomagnification, and it is a serious problem in the higher levels of food chains (Woodwell 1967, Oliver and Niimi 1985).

Since the transfer of energy from lower to higher food levels is extremely inefficient, biomagnification results from energy at each food level being produced from a much larger energy ingested from the level below. For instance, herbivores must eat large quantities of plant materials (which may be contaminated with toxins), carnivores must eat many more herbivores, and so on. Because fat-soluble toxins are not excreted, the carnivore gets accumulated dosages of toxins from its food over a long period of time. Thus, in the higher trophic levels of carnivores, including humans, these toxins have reached high concentrations. As a result of biomagnification in the food that they eat, many people have concentrated pollutants in their bodies. The powerful insecticide DDT, for example, has been found in mother's milk in the United States at 0.5 parts per million. This is 100 times the FDA's maximum permissible amount in human food.

Nonbiological disintegration of pollutants is complex, and several factors such as wind, water, and climate generally act together to achieve decomposition or decay. There seems little doubt that some pollutants, like carbon monoxide, react strongly with oxygen in the atmosphere and are rapidly converted to carbon dioxide. Moreover, radioactive isotopes by themselves will decompose to form harmless substances. The latter, of course, takes thousands of years.

During the past few decades, scientists have become increasingly aware that some pollutants may be degrading more slowly than expected. According to the Council on Environmental Quality (1985):

> One important aspect of the ozone and global warming issues is that the atmospheric lifetimes of gases such as nitrous oxide, and [CFCs], are known to be very long (as much as 180 years). Consequently, if there is a change in atmospheric ozone or climate caused by increasing atmospheric concentrations of these gases the full recovery of the system will take several tens to hundreds of years after the emission of these gases into the atmosphere is terminated.

Environmental Damage and Legal Rights of Natural Objects

Based upon what they affect, we can group pollutants into two categories: biocides (from bio = life and cide = kill) and locucides (from locus = place). Biocidic pollutants harm organisms directly, because they harm specific tissues or entire organisms. From the ecological viewpoint, locucidic pollutants are more dangerous because they have an adverse effect on the natural environment or habitat of species.

A well-known example of a biocidic pollutant is DDT, now banned in the United States. Too much DDT in a bird's tissue interferes with the deposition of calcium in its eggshells. As a result, birds lay eggs with shells that are thinner than normal. These fragile eggs break easily allowing no protection for the developing chick inside the egg. Predatory birds, which are never in abundance because they occupy the higher food levels, are the ones most seriously affected.

Locucidic pollutants also have the potential of affecting the ecosphere in both uncertain and dramatic ways. The global ecosphere is being influenced by locucidic pollutants such as discharges of chemicals (CFCs, carbon dioxide, and other gases), deforestation, and other habitat modification. As described before, both the greenhouse warming and the possible ozone depletion are examples of the ways in which locucidic pollutants may be inadvertently changing the ecosphere. Because the various parts of the ecosphere interact in innumerable ways on scales ranging from the molecular to the global, this presents enormous scientific uncertainties. Thus, locucidic pollutants are changing the ecosphere on a grand scale, the outcome of which is still in question, and the consequences of which may be far-reaching indeed.

It is widely recognized that the greatest threat to biological diversity is that of habitat modification. Construction, urbanization, and other developments, such as agriculture, and intensive management for one or a few species, contribute to the decline of available habitats. As will be mentioned in chapter 3, prime forest land is being lost rapidly to expanding urban areas and farms. Mankind is changing or destroying the natural habitat too rapidly for most species to adapt, resulting in the extinction of thousands of animal and plant species. The importance of maintaining biological diversity and the legal rights of natural objects will be summarized in the following paragraphs.

In ecosystems, organisms are challenged by fluctuations in the physical environment, predation, parasitism, and competition for resources. Extinction results when species highly adapted to one set of conditions are unable to survive under new conditions. The history of the dinosaurs attests to the ultimate fate of many groups of organisms—extinction. In fact, scientists estimate that 99 percent of all species that have ever existed since the origin of life three to four billion years ago are now extinct (Raup 1986). These species were not able to adapt to changes in their environments.

From 1600 A.D. to the present, extinction has been proceeding at a frightening pace. Approximately 150 mammals and birds have become extinct and gone the way of the dodo. Of the five to ten billion species in existence, at least one million are likely to be lost within our lifetime. According to a National Research Council report of tropical biology, more than half of all existing species could cease to exist by the year 2100.

The underlying principal causes of extinction are the expansion of human populations, along with trophy hunting, economic harvesting, deforestation, wetland drainage, urbanization, agricultural clearing, and biocidic pollutants. Humans are changing the environment and destroying the natural habitats too rapidly for most species to adapt.

In view of their ecologic role in ecosystems, the impact of such extinction might be devastating. The rich diversity of species and the ecosystems that support them are intimately connected to the long-term survival of humankind (Ehrenfeld 1976, Josephson 1982). By reducing biological diversity, humanity is squandering its greatest renewable natural resource, on which we depend for nutrients, energy, fibers, wood, wood preservatives, drugs, beverages, gums and related substances, essential oils, resins, tannins, cork, dyes, fatty oils and related substances, latex products, aesthetics, and countless other benefits (Ruggieri 1976).

An endangered species, such as the sea otter, may have a significant role in its community. Although it is relatively scarce, the organism controls the structure and functioning of the community by its activity. The sea otter, for example, in relation to its size, is perhaps the most voracious of all marine mammals. The otter feeds on sea mollusks, sea urchins, crabs, and fish. It needs to eat more than 20 percent of its weight every day to provide the necessary energy to maintain its body temperature in a cold marine habitat. The extinction of such keystone species from the ecosystem would cause great damage.

Another reason the irretrievable loss of species must be slowed down is for the preservation of future natural ecosystems and human life-support systems. Traditionally, species have always evolved along with their changing environment. As disease organisms evolve, other organisms may evolve chemical defense mechanisms which confer disease resistance. As the weather becomes drier, for example, plants may develop smaller, thicker leaves, which lose water more slowly. The environment, however, is now developing and changing rapidly, but evolution is slow, requiring hundreds of thousands of years. If species are allowed to become extinct, the total biological diversity on Earth will be greatly reduced. Therefore, the potential for natural adaptation and change will also be reduced, and diversity of future natural ecosystems and human life-support systems may be endangered. Humankind's well-being depends on the preservation of as many species as possible. The president of the World Wildlife Fund, Russell Train (1978), commenting on this threat, wrote:

I have spent most of my time over the past several years working on a variety of pollution problems—air, water, and chemical among others. As I review these efforts, I am struck by the fact that the real "bottom line" is the maintenance of life on this earth. Time is running out rapidly on the natural systems of the earth, and particularly on the survival of species. The loss of genetic diversity which threatens everywhere and the resulting biological impoverishment of the planet have grave implications for our long-term future.

We need nothing less than a comprehensive program worldwide to preserve and protect representative ecosystems.

The recent past has demonstrated increasing consciousness on the part of some people concerning the environment and conservation of species (Myers 1980, Fitzgerald 1986, Spitler 1987, Wolf 1988). Some laws have been enacted to control trafficking of endangered species and products, such as exotic birds, rhinoceros horns, elephant ivory, reptile leather, and cacti. Migratory wild animals do not recognize man's political boundaries. They migrate from one nation to another as well as through international zones. These international travellers need international laws protecting them as well as their habitats. Survival of each species and ecosystem threatened by human intervention should be assured. International laws must provide these natural objects with statutory or constitutional protection. The legal system, for example, can guarantee that species and ecosystems have natural rights which grow out of the nature of the universe and depend on their structure and function, as distinguished from those that are created by law and depend upon civilized society, such as civil and political rights. If that is the case, then the holder of this natural right would have to require: (1) some amount of review to actions that are inconsistent with that natural "right"; (2) the natural object institute legal actions at its behest; (3) the court or legal authority take injury to the natural object into account in determining the granting of legal relief; and (4) relief run to the benefit of the natural object. In a book significantly entitled *Should Trees Have Standing?* Christopher D. Stone (1974) writes:

> Now, to say that the natural environment should have rights is not to say anything as silly as that no one should be allowed to cut down a tree. We say human beings have rights, but—at least as of the time of this writing—they can be executed. Corporations have rights, but they cannot plead the fifth amendment; In re Gault gave 15-year-olds certain rights in juvenile proceedings, but it did not give them the right to vote. Thus, to say that the environment should have rights is not to say that it should have every right we can imagine, or even the same body of rights as human beings have. Nor is it to say that everything in the environment should have the same rights as every thing in the environment.

In order to recover legal damages for injury to natural species and ecosystems, scientists could calculate the physical and biological requirements needed to maintain the structure and function of species or natural ecosystems. Economist can then determine how much money is needed to preserve the structure and function of the natural object. This includes all damages. The money could be used by the legal guardian of the natural object to decontaminate the ecosystem or species habitat, restock the ecosystem with its species, purchase property to restore the original size of the ecosystem, etc.

One new concept called natural-area zoning can be used to protect prime natural ecosystems, such as aquatic and terrestrial biomes. To be really effective, international laws must provide natural biomes with statutory or constitutional protection. Worldwide efforts to protect natural biomes are being made under the auspices of many governmental and other institutional organizations (World Wildlife Fund, Nature Conservancy, National Audubon Society, New York Zoological Society, Sierra Club, Environmental Defense Fund, Center for Environmental Education, etc.). These public and private organizations have a variety of options to explore in their efforts to preserve prime natural biomes. Natural-area zoning can be used to protect wetlands, coastal dunes, wild rivers, game preserves and other natural habitats, forest, fish spawning areas, and other vital areas.

Other current biome-conservation issues of international concern include: (1) conservation of representative large ecosystem reserves and natural sites of world significance; (2) maintenance of regional wildlife management and conservation training in institutions, such as the College of African Wildlife in Tanzania; and (3) research in the minimal critical size of ecosystems required to sustain biota, which is currently taking place in Brazil.

Pollution Response and Scientific Uncertainty

Another important aspect of the effect of pollution is the dose required to cause environmental damage. Pollution response, as herein defined, is the change in the effect of a pollutant in response to a change in its concentration. Generally, three types of pollution response are distinguished: the linear assumption, the greater-than-linear assumption, and the threshold assumption. In the linear effect, the environmental damage rises linearly with concentration (figure 2-5). This linear assumption implies that the total damage or risk is directly proportional to the total accumulated exposure. Generally, the assumption is also made that the risk is the same regardless of the rate at which the dose is delivered. This curve has generally been found to occur with the radioactive substances, mercury, lead, cadmium, and asbestos.

In the greater-than-linear effect, as pollutant concentration increases, the environmental damage also increases but at a de-

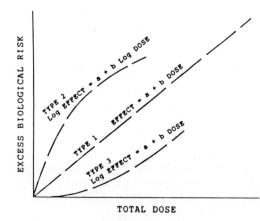

Figure 2-5. Three possible relationships between excess environmental risk and pollution doses at different exposure levels.

creasing rate. Hence, there is a decreasing rate of damage as its concentration increases. A greater-than-linear hypothesis implies that damage induction per total pollution dose would be greater at low exposures than at high exposures. This type has been shown to occur with thermal pollutants.

In the third type of response, there is a threshold effect as pollutant concentration increases. That is, this type produces no effect until the threshold is crossed. Generally, the assumption is also made that so long as a given threshold is not exceeded, the damage from pollution would be completely repaired as quickly as it is produced; therefore, there would be no harmful result. This curve has generally been found with biodegradable pollutants.

It is important to note that although each pollutant in the environment may not be present in proportions that cause environmental damage, interaction effects of pollutants are bound to occur. The additive effects of many toxic pollutants in low concentrations can put significant stress on the global ecosphere. Each pollutant increases the range of factors to which organisms must respond. When combined, some pollutants exert a greater influence than the sum of their individual effects (synergistic effect). The synergy between asbestos exposure and smoking in causing lung cancer, is a well-known example of a synergistic effect (Surgeon General's

Report 1988). On the other hand, some toxic substances may have antagonistic effects which could reduce the overall toxicity from the added effects. As alluded to already, it is simplistic to deal with individual pollutants as isolated from the rest of the system. For example, the chemicals (i.e., CFCs, carbon dioxide, methane, and other gases) which are expected to deplete the ozone layer, are the same chemicals which are expected to warm the ecosphere. Moreover, increasing concentration of methane is also expected to increase ozone in the atmosphere and may contribute to the forest damage that is occurring. (We shall take up the problem of scientific uncertainty and pollution response in chapter 5.)

STABILITY OF HUMAN SYSTEMS AND CARRYING CAPACITY

In the first section of this chapter, the anthroposystem theory was postulated in an attempt to describe any environmental system developed by humankind that can perpetuate itself. Modern science is based on quantification of theory, and indeed, precise numerical terms or measurements and scientific testing are almost inseparable. Only after a system is quantified can one mathematically predict the system's stability. This section identifies and quantifies the position or role in the environment that an anthroposystem occupies. The quantification of the anthroposystem is one of the most important aspects of the study of humans and their environment. With such knowledge, the design of a stable human system and an interrelated long-range life-support system can be undertaken.

Types of Stability

As we have seen, a real anthroposystem is an open system; that is, there is a constant flow of resources between it and the external environment. If water enters at exactly the same rate as it leaves, the amount of water in the system remains constant despite the fact that it is continually being renewed. We would say that water is in a steady state or equilibrium, meaning that its concentration in the system does not vary, even though it is continually being exchanged with the external environment. When the flow of the same type of resources into an anthroposystem is equal to the outflow of the same type of resources, the system is at a state of equilibrium for those resources. That is, $F_{01} = F_{14}$ (see figure 2-3). The resource content of the system may oscillate somewhat above and below its carrying capacity because the inputs may not have an immediate impact on outputs (figure 2-6).

In order for an anthroposystem (or nation) to maintain its integrity (its environmental structure and function), it must maintain a position of equilibrium on a resource treadmill. That is, integrity is made possible only through the continuous acquisition of matter and energy in the face of changing external and internal conditions. This tendency to regulate the environment is called homeostasis. In a stable natural ecosystem, for example, the number of prey and predators being born is approximately equal to the number dying, resulting in their populations remaining at a steady state. The principle of homeostasis, as it applies to natural ecosystems, is popularly referred to as the "balance of nature." Recently, the term "Gaia" has been postulated to describe the Earth's entire ecosphere as a single living organism (Lovelock 1979). Gaia's

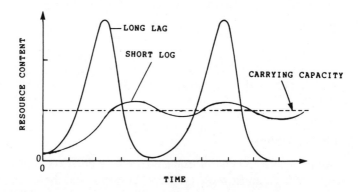

Figure 2-6. Carrying capacity and anthroposystem size (resource content). For any given system in an area, there is a carrying capacity or optimum size that the area can indefinitely support with its ability to produce and recover resources for that system. Usually, actual system size fluctuates closely around the theoretical carrying capacity. However, as this figure shows, the longer the time lag between inputs and outputs, the greater the oscillations.

integrity is believed to be controlled by a worldwide homeostatic mechanism like the balance that apparently exists between respiration and photosynthesis. The way in which homeostasis may operate in anthroposystems is the subject of this unit. Homeostasis involves regulatory mechanisms termed congeneric and feedback/feedforward.

Congeneric Homotaxis. "Congeneric homotaxis" is a term originally coined by Hill and Durham (1978) and Hill and Wiegert (1980) to describe a stability which is essentially a function of the redundancy in functional parts of the system. For example, if several car manufacturing corporations are present, each with a different resource requirement, the rate of production of cars within the system as a whole can remain stable despite changes in resource input. Congeneric homotaxis is correlated to the ability of the system to adapt to changing resources. In turn, the ability of a system to adapt to changing resources depends on the tolerance ranges of the resource and their relative importance.

Logically, the tolerance range of an anthroposystem to a given resource affects its ability to maintain itself in a particular environment. For example, let us imagine an island anthroposystem or nation. This is perhaps the easiest system to analyze because it has a clear geographical boundary. Suppose further, for the sake of analysis, that our system requires only five factors: temperature, oxygen, precipitation, chromium, and petroleum. For each factor, the system will function most efficiently within a fixed range, as indicated in figure 2-7. We can present the first two resources, temperature and oxygen, as the axes of a two-dimensional graph on which any single point represents a locality with a specific temperature and oxygen level. As shown in figure 2-8, the shaded area of the graph indicates all the possible temperature-oxygen levels within which the system can maintain its integrity, or structure and function.

We can add third, fourth, and fifth axes representing precipitation, chromium, and petroleum requirements. The optimum levels pass through the origin (0,0) of the cartesian coordinates, thus delineating a 5-axis space, indicating tolerable combinations of the five variables. The two-dimensional area that results is a geometric representation of that system's configuration (figure 2-9).

Each given factor can also be visualized as one coordinate in an infinite dimensional space with upper and lower limits for the

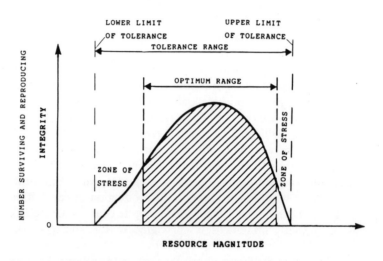

Figure 2-7. Effects of resource magnitude on an anthroposystem's integrity. Integrity declines as availability of any factor deviates from the optimum level. This occurs as a result of the accumulation of pollutants and/or because of the law of diminishing returns. Generally, especially in developed countries, the integrity is affected by pollution and the unintended by-products of industrialization rather than by the depletion of resources (Daly 1973, Valaskakis 1981). Habitat factors (for example, temperature, space) also have their minimum, optimum, and maximum levels.

maintenance of the integrity indicated. The many dimensions of this "hypervolume" represent environmental parameters on which one conceptually plots the system's limits. The hypervolume concept has become popular in describing the ecological niche of species (Hutchinson 1958). By analogy one can state that the configuration and hypervolume of an anthroposystem represents its environmental niche.

The quantification of an anthroposystem's configuration (and hypervolume) aids in understanding its environmental stability. The simplest method of expressing a system's congeneric stability is in a geometric index such as presented in figure 2-10. One of the basic problems with the use of a geometric index is the determination of the relationship between magnitude and resource ranges, and their position on the absolute scale of the cartesian coordinate. Also, interactions among the factors almost always occur, and this would

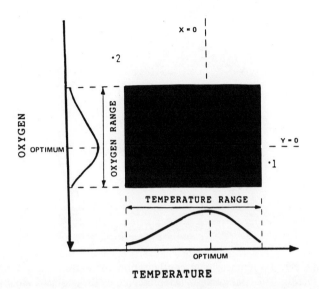

Figure 2-8. Two dimensions of the configuration of an anthroposystem. Any system's integrity is restricted to a range of temperature and oxygen levels. Each point in the shaded area represents a possible environment, a combination of temperature and oxygen levels. If a given area in the universe represents only environments in which the temperature and oxygen levels fall outside the shaded area, as do points 1 and 2, no proposed system located there could maintain its integrity.

alter the configuration of the system. Thus, the actual geometric stability index will have to await the completion of monographic research.

Another index of congeneric stability could be based on the importance of each factor. That is, not all factors are equally important in maintaining the integrity of a system. Of many factors required by a system, relatively few exert a controlling influence by virtue of their qualitative and quantitative roles. For example, the lack of petroleum in a petroleum-based industrialized nation would create an immediate economic and political crisis, whereas a nation that has its energy demands distributed among various resources is less vulnerable and more stable. (A system which relies equally on all its resources would be more stable than a system which relies heavily on one particular resource.) The degree of

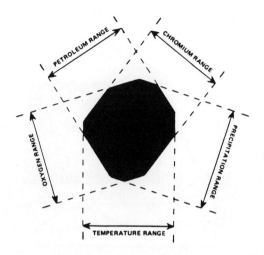

Figure 2-9. Two-dimensional configuration for a system requiring only five factors. The configuration is represented by the perimeter (solid) lines. If a space contains combinations of the five variables that lie within the range indicated by the shaded area, a system at that location could maintain its integrity. For illustrative convenience, the five axes meet at the 0,0 coordinate.

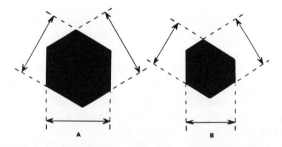

Figure 2-10. The shaded area that represents an anthroposystem serves as a numerical measure of the limits within which that system can function. It is intuitively obvious that, for given factors, a system with a greater magnitude and thus greater geometric area (diagram A) will be more stable than one with a smaller magnitude (diagram B). The former will have a greater ability to maintain its overall integrity in the face of changing factors.

necessity of a specific factor to the maintenance of a system's integrity can, therefore, be expressed by an appropriate index of importance that sums up each factor's relative value in relation to the anthroposystem as a whole (Santos 1983).

The geometric and importance indices can be viewed as examples of the so-called "non-oscillation stability" or "stability resilience" (Hill 1975). Stability is essentially a function of the complexity of the recycling of matter or energy flow, in which a large number of interacting pathways provide mechanisms for adjustment of stress. Presently, ecologists are not in agreement as to how complexity contributes to stability in natural ecosystems (Begon et al. 1986).

Limiting factors are those factors that exert some restraining influence upon a system through incompatibility with its requirements. Whenever any of the variables goes below a fixed fundamental minimum or are limiting, the system will not survive even though all other variables are at satisfactory levels (point 2, figure 2-8). Similarly, if any of the variables goes above a fixed fundamental maximum limit, the system will not survive even though the other variables are at satisfactory levels (point 1, figure 2-8). The limits of tolerance thus occur at both maximum and minimum points, and the range within these limits is the area of survival. Therefore, whenever any single factor becomes limiting, either at its minimum or maximum limits, the system cannot survive even though all other factors are at a satisfactory level. This principle was first elaborated by Justin Liebig while studying organisms and has been called Liebig's Law of the Minimum. According to this ecological law, "when some process depends on several different factors, the speed of the process at a given time is limited by the 'slowest' factor. The slowest or limiting factor may be either too little or too much of something" (Brewer 1988).

Feedback/Feedforward Control. In addition to congeneric homotaxis, feedback/feedforward also contributes to stability. Feedback refers to the return of some of the output of a system as input. It occurs whenever a response to a stimulus feeds back to alter the original stimulus. If the output augments the system, the feedback is positive; if it tends to inhibit the system, the feedback is negative. Organisms usually keep conditions optimum within their bodies through negative feedback control. Among the many examples of feedback regulation of biological processes is the automatic maintenance of our internal body functions by the nervous and endocrine systems. Another example is the ordinary household thermostat. The thermostat is set to the desired temperature. A drop in room temperature turns on the thermostat, sending a signal to the furnace so that it produces heat. The heat output of the furnace raises the room temperature, which turns off the thermostat, shutting down the furnace. This is called "negative" feedback because the result (a higher temperature) is opposite the initial stimulus (a low temperature). Over time, the amount of heat produced is balanced with the rate of heat loss minimizing the difference between actual and optimum levels.

We can also note examples of feedforward mechanisms when sensory cells in human skin detect a drop in air temperature. They send signals to the brain to "expect" a modification in blood temperature. The brain sends an impulse to metabolic and muscular systems that can function in raising body temperature. With feedforward mechanisms, corrective measures can sometimes commence even before the external environment significantly changes the internal environment.

Feedback/feedforward control mechanisms demonstrate oscillation when, for any reason, there is a lag in response. The response then takes longer than it should to reach equilibrium, and the system overshoots or undershoots. Feedback/feedforward then occurs in the opposite direction. As demonstrated in figure 2-6, oscillation of this type accounts for much of the normal balance that occurs in systems. To minimize oscillations, the output should be proportional to the error signal, which is the difference between optimum point and actual resource content. That is, if the resource content is below the optimum carrying capacity (optimum point) by only a small amount, the control center would be controlled so that the output would be low. For a greater deviation from the optimum point, the controlled output of the controlled center would be increased in direct proportion.

Just as feedback/feedforward mechanisms regulate systems, an anthroposystem or nation can minimize environmental oscillations. Moreover, since we have the capacity to anticipate and adapt to events before they happen, we can modify goals and attempt to balance resource gain and resource loss in order to keep the resource at optimum level.

The mechanism described here to control a complex human-environment system is an example of what is called "adaptive feed-back—feedforward control systems" or "universal adaptive systems." According to Bennett and Chorley (1978):

These control systems are capable of using past behavior to modify future response and/or the set point performance measure. The internal model is capable of reprogramming itself and must employ additionally some feedback elements especially the use of the fed back error signal (Rosen 1975). Additionally the system has set of "effectors" (Kuhn 1974) with which it can manipulate the environment and create new sensory modalities for itself by fabricating instrumentation and sensors to gain new knowledge about environmental behaviour.

A universal control mechanism in an anthroposystem can be represented by the block diagram in figure 2-11. The controlling center contains two subunits, an error detector and a controller. The output signal is sampled by a detector which generates a feedback signal that is transmitted by way of the feedback loop to the error detector. The error detector subtracts the output of the system (the feedback signal) from the set-point or optimum level, and thus generates an error signal. The second subunit of the controlling center is an adaptive controller, which receives the error signal as its input, and generates a controlling signal as its output. The controlling signal is the input for the controlled center; it continually alters the output of the controlled center in a direction that minimizes the error signal. Moreover, the adaptive controller anticipates human behavior and environmental disturbances, and, before they have had time to affect stability, a modification in the controlling signal is made to counteract future problems.

The adaptive controller would function much like Spinoza's free intuitive-type executive. As Churchman (1971) redefined the free intuitive executive system, it "is based on intuition leading to knowledge which arises through the essence of the object perceived, such that the executive has a valid theory to explain why knowledge (i.e., inputs into the inquiring system) occurs. In such a system the executive has a key role in filtering and organizing the information inputs from the real world." We will encounter the concept of the universal control mechanism in chapter 5, when we discuss a proposed World Environmental Authority.

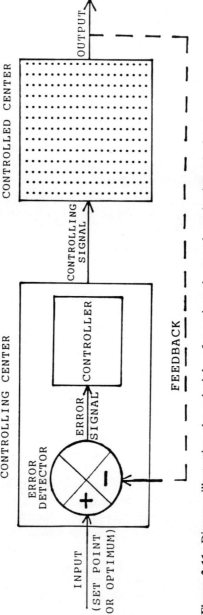

Figure 2-11. Diagram illustrating the principle of a universal control mechanism. A detector responds to a stimulus, such as high output, and signals an adaptive controller that directs an adaptive response, such as the reduction of output. Once normalcy, such as optimum resource, is achieved, the detector is no longer stimulated. (Simplified from Bennet and Chorley 1978)

Resources and Human Civilizations

In assessing an anthroposystem's stability and carrying capacity, we are describing its survival ability in terms of its predictable surroundings or environment. However, a complete description of an anthroposystem's total configuration would have to include other criteria, such as the possibility of natural catastrophes (e.g., earthquakes), as well as political and economic factors. Probabilistic predictions of natural catastrophes can be made and are based on recurrence periods of rare events (Gumbel 1941, Schleidegger 1975). Political and economic systems are goal oriented, and thus no assessment of their effect on the anthroposystem's configuration can be completely free of social influences. Or, as Gerlach and Hine (1973) express, "Human systems . . . are different qualitatively from animal organic systems in their capacity for self-awareness, symbolization, and rational thought." Social systems, however, function within the ecosphere and as such can be viewed in ecological context (Santos 1974). No social system, for example, can go against the second law of thermodynamics. As Clapham (1981) wrote:

A democratic society can choose how to manage its landscape, but it cannot determine the environmental principles that govern the responses of the landscape to that environment. Nor can it choose how others manage theirs. For example, we can choose to have wilderness or to cut virgin trees, but we cannot choose that soil will not erode once the trees are cut. We can choose to have cities that dump toxic wastes into rivers and estuaries, but we cannot choose to have salmon still swim through the polluted water to reproduce.

Historians and scientists are beginning to appreciate the importance of resources to civilizations. Recent evidence suggests that the rise and fall of civilizations can be largely attributed to environmental factors (Jacobsen and Adams 1958, Lowe 1985, Hammond 1986). As restated by Watt (1982), the decline of many civilizations can largely be explained "in terms of the intensity of pressure by the civilization on the resource base that supports it," and by the attitudes of the inhabitants towards "the importance of wise management of the resource base, so as to make it last over the long term.'' The same phenomenon of ecological deterioration has been discussed by Charles H. Southwick (1985), who pointed out:

> The span of history from 5000 B.C. to 200 A.D., which we know primarily as the period of great civilizations—Sumeria, Babylonia, Assyria, Phoenicia, Egypt, Greece, and Rome—was also a period of unprecedented environmental disturbance. We tend to concentrate our attention on the superb achievements of these civilizations in literature, art, government, and science, while we virtually forget their incompetence in land management. These golden civilizations prospered at the expense of their environments. They left a landscape which has never recovered, and a legacy to future civilizations which ushered in period of dark ages lasting for more than a thousand years.

Thus, ecological forces are more important in regulating past, present, and future history than one would expect. This "surprising" behavior of natural systems has been labelled "counter-intuitive" (Forrester 1973).

We have been realizing, recently, that human activity may be again inadvertently and irreversibly deteriorating its matrix. In the following chapter, we will consider the criteria for determining the earth's carrying capacity for humans.

3   *Criteria for Determining Earth's Carrying Capacity*

The human growth curve in figure 1-2 certainly shows a long lag phase and a recent sharp turn upward into the exponential growth phase. Perpetual growth is impossible, at least as long as humans remain on earth. A key question that remains is how we can determine the carrying capacity of the earth for humans. Also, which laws, principles, models, or criteria should be employed by mankind in determining the carrying capacity. The easiest path which demands little technological know-how and decision-making is that our society reduce its population/pollution to such a low number that the question "What is the carrying capacity?" becomes moot. If we reduce our numbers so that we become a part of the ecological steady state, our population and pollution would have an insignificant impact on the natural world. The expanding human population, worldwide famine, and pollution, however, make the issue of determining the carrying capacity of the earth a crucial one.

The optimum world population, as discussed in previous chapters, in that which is the most favorable level for the maintenance of stability on our planet. Arriving at the figure for any given period involves considering the sum of the demands that individual nations make on the environment and how these demands can be met without degrading or destroying that nation's ecosystems—in other words, estimating the population/pollution of each nation at the optimum carrying capacity of its environment.

S. Fred Singer (1971) expressed this concept in more general terms:

It is clear that optimum is not synonymous with maximum; in fact, the optimum level should be well below the maximum sustainable population level. This is sharp contradistinction to, say, the raising of cattle where the optimum, from the rancher's point of view, means the level which will result in a maximum financial yield.

No one knows exactly what the world's ultimate carrying capacity is and under what circumstances we will approach it. However, eventually we will reach the carrying capacity of the earth for humans. Whether our population will level off at the carrying capacity and follow an S-shaped curve (figure 1-4), or experience a dramatic sudden decrease and follow the J-shaped pattern (figure 1-5), or a combination thereof is not clear. A J-shaped population growth curve would be the most tragic for our society.

The preceding not withstanding, one thing is quite certain: no new medical care or technological advances can keep pace with that growth curve to supply resources and assimilate all wastes for all those people while leaving all the world's ecosystems intact. The only certainty is that we must voluntarily control population and pollution or the death phase will eventually control it for us. Lester R. Brown (1987), project director of the Worldwatch Institute, eloquently and concisely expressed the problem of the crash phase:

Once populations expand to the point where their demands begin to exceed the sustainable yield of local forests, grasslands, croplands, or aquifers, they begin directly or indirectly to consume the resource base itself. Forests and grasslands disappear, soils erode, land productivity declines, water tables fall, or wells go dry. This in turn reduces food production and incomes, triggering a downward spiral.

In analyzing the concern of human fertility and overpopulation, Potter (1988) proposed a generalized sequence of events that may cause mankind to overshoot the carrying capacity (point of no return):

1. Carrying capacity is increased by technology.

2. Population increases by medical technology and environmental technology.

3. Governments and private multi-national corporations are unable to manage ethical problems of distribution and equity.

4. Political unrest escalates among competing ethnic and religious groups.

5. Military solutions through exploitation and appropriation of science and technology are sought.

6. Development of science and technology for the common good withers because of competition with military demands.

7. Carrying capacity relative to human needs begins to enter [the] crash phase.

8. Widespread lowering of lifespan results.

9. Miserable survival remains as the outcome.

There are two main criteria for determining the optimum carrying capacity: socioeconomic and ecological. In this chapter we will explore these two criteria, examine their weaknesses, and evaluate their practicality.

SOCIOECONOMIC CRITERIA

The most common standards or criteria by which a government determines the number of people in its territory are based on aesthetics, economics, sociology, politics, or other human concepts. For example, population growth may be encouraged by an expansionist dictator for the sake of maintaining a larger army even though so large a population may be ecologically unsound. Population may also be fostered for the purpose of a larger group of artisans, traders, and taxpayers. Presently, most economists believe that our society relies on a constantly expanding investment in business, a continuous flow of goods and services, and quantities of consumers in order to maintain existence. They contend that these criteria are essential in order to make jobs available for the growing population and so that incomes will stay in line with inflation. It is certainly the only way investment money can remain earning high interest rates. The economist Julian Simon (1984), for example, contends that "Adding people causes problems, but people are also the means to solve these problems." According to him, (1) people are the earth's most vital resources; (2) the concerns of resource depletion and pollution can be solved by trained minds and imagination; and (3) more people generate higher consumption and thus spur economic development. Simon's vision corresponds in many respects to those of the U.S. Administration's policy statement on economic growth and population development issued at the U.N. International Conference of Population (see chapter 1).

Another economic criterion based on per capita income has also been postulated. For example, if a nation has 100 million inhabitants and per capita income is $25,000, it is projected that if it increases its population to 150 million people its per capita income would be $40,000; however, if it increases to 200 million people the per capita income is projected to be $27,000. Under this criterion, the most desirable population would lie at its level of the maximum per capita income or 150 million people.

Many issues are raised by socioeconomic criteria. Some, typically those based on aesthetics or "quality of life," are difficult to quantify. Consequently, it has been proposed, carrying capacity ultimately depends on society's value judgments (Wagar 1964, McHale and McHale 1976, Shelby and Heberlein 1984). The quest for determining the carrying capacity using quality of life criteria was explicit in the hearing before the U.S. Congress on the effects of population on natural resources and the environment:

> I define an optimum population as one that is large enough to provide the diversity, leisure, and substance whereby the creative genius of man can focus on the satisfactory management of his ecosystem, but not so large as to strain the capabilities of the earth to provide adequate diet, industrial raw materials, pure air and water, and attractive living and recreational space for all men, everywhere, into the infinite future. I take it as self-evident under this definition that there is an optimum population, or at least optimal limits. (Cloud 1969)

Apparently, applying the concept of carrying capacity to humans is not easy. Even more difficult is applying the concept when human values are involved. However, just because the concept is difficult to quantify does not mean that it has no significance. William R. Catton, Jr. (1987), professor of sociology, captured the essence of this point when he stated:

Political and economic leaders, and social scientists, tend to exaggerate any recognition that carrying capacity is not constant into the supposition that it is infinite. The fact that carrying capacities can be difficult to measure cannot exempt populations from the consequences of exceeding their environments' power to sustain them. The human prospect would be brighter if somehow these points were to be central to the agenda for the next superpower summit meeting—but of course they won't even be mentioned.

Catton calls attention to the concept of equity (subjective) and sustainability (objective) as two distinct dimensions of the carrying capacity theory. He argues that equity or socioeconomic concerns have led some politicians and industrialists to state that carrying capacity is a meaningless theory. As he so aptly put it:

From these comparisons what is most vital to note is that the question of a load's sustainable magnitude is an objective ecological issue, not a value question. The question of which trade-off is preferable to its alternative is a value issue. Choosing whether to increase the user population at the cost of lowering its standard of living or to raise affluence at the cost of population reduction depends on value judgment. But it is a serious mistake to suppose this denudes carrying capacity of any objective meaning. (Catton, Jr. 1987).

Catton's views correspond in many respects to the equation given in chapter 1, where it was stated that the environmental impact is measured not only in terms of the population size but also in terms of the resources used and the pollution caused by each individual in the population. Catton went on to suggest that "Those who aspire to leadership positions should be required to demonstrate that they understand the load and carrying capacity concepts."

Another critical problem with socioeconomic criteria is that they assume that society will continue to grow. However, as most economists recognize, beyond a certain point, the continual increase of population may lead, and in many countries has already led, to a situation of diminishing returns. In addition, per capita income may not be a useful indicator of economic well-being. As a 1972 report, the President's Commission on Population Growth and the American Future, points out:

Rapid population growth will cause more rapid growth in the size of the economy, and correspondingly greater demands on resources and the environment. People will not be better off economically with more rapid population growth . . . income per person is higher under the slower growth assumption. Rather, increases in the number of people simply multiply the volume of goods and services produced and consumed.

Perhaps the most fundamental flaw in using socioeconomic criteria is the fact that they make ecological criteria subunits of human values (Santos 1974). As the economist Herman E. Daly (1974) has explained, "Our economy is a subsystem of the earth, and the earth is apparently a steady-state open system. The subsystem cannot grow beyond the frontier of the total system and, if it is not to disrupt the functioning of the latter, must at some earlier point conform to the steady-state mode." In a more recent article Daly (1986) elaborates and indicates that "as the economy grows beyond its present scale, it may increase costs faster than benefits and initiate an era of uneconomic growth which impoverishes rather than enriches."

One of the first economists to propose that our economy should become a steady state economy was Kenneth E. Boulding (1966). To explain his reasoning for a transition to a steady state economy, he constructed an analogy of the planet Earth as a spaceship (a closed system) to denote the finite limitations upon the resources of the earth and the futility of consuming those resources. Boulding suggested that:

The closed earth of the future requires economic principles which are somewhat different for those of the open earth of the past. . . . The closed economy of the future might . . . be called the "spaceman economy," in which the earth has become a single spaceship, without unlimited reservoirs of anything, either extraction or for pollution, and in which, therefore, man must find his place in a cyclical ecological system which is capable of continuous reproduction of material form even though it cannot escape inputs of energy. The difference between the types of economy becomes most apparent in the attitude towards consumption. In the cowboy economy, consumption is regarded as a good thing and production likewise; and the success of the economy is measured by the amount of the throughput from the "factors of production," a part of which, at any rate, is extracted from the reservoirs of raw materials and noneconomic objects, and another part of which is output into the reservoirs of pollution. If there are infinite reservoirs from which material can be obtained and into which effluvia can be deposited, then the throughput is at least a plausible measure of the success of the economy. The Gross National Product is a rough measure of this total throughput. . . .

By contrast, in the spaceman economy, throughput is by no means a desiratum, and is indeed to be regarded as something to be minimized rather than maximized. The essential measure of the success of the economy is not production and consumption at all, but the nature, extent, quality, and complexity of the total capital stock, including in this the state of the human bodies and minds included in the system. In the spaceman economy, what we are primarily concerned with is stock maintenance, and any technological change which results in the maintenance of a given total stock with a lessened throughput (that is, less production and consumption) is clearly a gain.

Inherent in the steady-state notion is that eventually the growth rate of our society will slow down and an equilibrium or steady state with the environment will be more or less achieved. Theoretically, at least, this last growth stage will be mature, self-maintaining, self-reproducing through development stages, and relatively permanent. Our society will be tolerant of the environmental conditions it has imposed upon itself. This terminal society will be characterized by an equilibrium between gross industrial and agricultural production and total consumption, between the energy captured and the energy released, and between the uptake of natural resources and recovery of resources by the decomposer component. In this final stage, more resources will be devoted to the maintenance of the complex human system than to the production of the system.

In the last stage of growth there will be a wide diversity and complex interaction of professionals along with well-developed spatial structure. This final stage will be the climax community of our society. The human system will be similar to an ideal anthroposystem. It will be a self-sustainable, functional, and structural unit of interwoven and overlapping hierarchies of organization which maintain civilization.

This proposed growth pattern of human society is very similar to what occurs in natural ecosystems; where the final stage of succession is reached or approached, the photosynthesis/respiration ratio approaches one. There is also greater biomass and diversity of species. The community is relatively more stable and self-sustainable. The body growth pattern of organisms is also similar. The rate of body growth, for instance, is very rapid at the beginning of an organisms's life, but it slows down in adolescence and becomes more stable in adulthood.

Apparently, the steady-state economy is inevitable, but the means of achieving it are largely unknown. There are many important and interesting publications on self-sustainable human systems. For more on self-sustainable theory, see publications by Boulding (1966, 1982), Ayres and Kneese (1971), Daly (1973), Brown (1982), and Brown et al. (1987, 1989).

ECOLOGICAL CRITERIA

Recognizing the difficulties inherent in using socioeconomic criteria, scientists have attempted to determine the carrying capacity of the Earth using ecological criteria. The exact number of people the world can support will vary based on the habitat or nutritional resources employed in the calculation (table 3-1). However, according to Liebig's Law of the Minimum, whenever any variable needed for survival falls below a fixed fundamental minimum or is seriously limiting, the population will not survive even though all other variables are at satisfactory levels. Therefore, optimum population is determined by the variable necessary for survival that is in the shortest supply (see figure 2-8). According to the American standard of living and total land criteria, the world is above the carrying capacity that an intensively managed global environment might hope to support with some degree of economic well-being and individual choice. Although it is difficult to gauge which ecological variable will be the ultimate limiting factor, the data indicate that the most important variables are total land, food, and heat buildup. Heat buildup and the "greenhouse effect" were summarized in chapter 2. Therefore, in the following pages, we shall review the interactions of food, land, and people.

Food and Carrying Capacity

From a historical point of view, food is probably the most useful indicator of supportable population. How important is food

Table 3.1
Estimates on Carrying Capacities of the Earth

LIMITING VARIABLE	POPULATION (in billions)	REFERENCES
American Standard of Living *	0.59–1.13	Watt (1982)
Total Land **	3.1	Westing (1981)
Food ***	10	National Acad. Sci. (1969)
Heat Buildup	15–20	Weinberg and Hammond (1972)
Oxygen	100	Campbell (1970)
Energy	100	Ibid.
Water	100	Ibid.

Note: Estimates on carrying capacities of the earth depend on the criteria used. Forrester (1982) appropriately points out that, "if one limit can be evaded, another will immediately be revealed."

*If the earth had the diet of the average American, used the level of fertilizer, and used as much space as the average American.
**Modeled after the level of an average of the world's nations of average wealth.
***All people are to be more than merely adequately nourished.

supply? We know from the whole course of human history that famines are forever causing troubles (Paarlberg 1988). Examples are legion: 1,828 famines in China from 108 B.C. to 1911; 201 in the British Isles; 2 famines in the Soviet Union, 1 from 1921 to 1923 and 1 from 1932 to 1934; and today famines in Africa and the Near East—just to name a few. According to Paarlberg (1988):

What share of human history is significantly affected by hunger and a dearth of food? It would take a bold man indeed to venture an answer. But one can say with confidence that over the centuries famine has been a major force limiting the numbers of people. The quest for additional food has outfitted armies and launched ships. Empires have risen and thrones toppled in the struggle to control food supply.
Saint John the Divine, writing from the Isle of Patmos around the close

of the first century of the Christian era, produced his visionary work, the Revelation, last book of the New Testament. In it he described the Four Horsemen of the Apocalypse: Pestilence, War, Famine, and Death. The Horsemen rode together. It would be difficult to judge which was the greatest scourge.

In the following paragraphs we will consider the major nutrients and symptoms of their deficiencies, and world food production.
Nutrients and Symptoms of Their Deficiencies. Various international agencies estimate that between five hundred million and one billion people do not consume enough food or are chronically undernourished. In fact, the world presently has more hungry people than it did when the eighties began (U.N. 1988). While some undernourished people reside in developed nations, most reside in underdeveloped nations. In particular, Africa and the Near East, with the fastest population growths ever reported, have had problems producing sufficient food for their populations.

It is estimated that one-half of the children in some developing regions do not get a balanced diet and suffer from malnourishment (malnutrition). These children are not getting enough required nutrients, especially protein, and suffer severely from malnutrition. In developing countries, approximately ten million children under the age of five starve to death each year, chiefly of malnutrition and the complications correlated to it. Possibly an additional ten to fifteen million older individuals die of starvation per year. Therefore, about 2,500 to 3,330 individuals starve to death each hour; many more individuals are on the brink of starvation, ill, or possibly irrevocably mentally retarded as a result of malnourishment and/or undernourishment when they were babies.

In humans, infant mortality is the most sensitive indicator of nutritional stress (Brown et al. 1989). For example, on the island of Madagascar, with eleven million people, infant mortality, is increasing by 3 percent per year. Madagascar's per person grain production reached an all-time high in 1967 and fell gradually until 1983, when the decline accelerated. Since then average grain consumption has declined by almost 20 percent, reducing food consumption below the survival level for many. Infant mortality increased from 7.5 to 13.3 percent between 1975 and 1985.

Nutrients provide the source of dietary energy and materials for the human body. The nutrient requirements fall into two broad categories: organic and inorganic compounds. The organic compounds mainly include carbohydrates, lipids, proteins, and vitamins. The carbohydrates (sugars, starches, glycogen, cellulose) are a principal source of energy. The oxidation within the body of 1 gram of carbohydrate in pure form yields about 4 kilocalories (Surgeon General's Report 1988). Starchy vegetable foods, such as bread and potatoes, provide the largest amounts of carbohydrates, but so do meats and seafood because they contain glycogen. Glycogen (animal starch) serves as a food reserve in humans and other organisms (other animals, bacteria and fungi); it can be broken down readily into glucose. Carbohydrates are also used as structural materials or as part of other molecules.

The lipids (fats, oils, waxes, steroids, carotenes) contain more than twice as much energy as carbohydrates, about 9 kilocalories per gram. The lipids, in addition to serving as a source of energy, function as important components of cells, energy storers, and hormones. Food rich in lipids include butter, margarine, meat, eggs, milk, nuts, and a variety of vegetable oils.

Proteins provide the physical framework of the human body. For example, the cartilage that binds bone joints together is composed of a protein (collagen). Other proteins serve to transport oxygen (hemoglobin). Others function as catalysts and regulate the speed at which chemical reactions occur in the body (enzymes), and still others as hormones and chemical messengers. Proteins are present in meat, fish, poultry, cheese, nuts, milk, eggs, cereals, beans, and peas. Proteins are built of long chains of basic molecular units called amino acids. There are twenty different amino acids which, except for eight, the human body has no difficulty in changing from one type to another. Those that cannot be changed or synthesized are called essential amino acids (they are lysine, tryptophan, threonine, methionine, phenylalanine, leucine, isoleucine, and valine) and must be provided in the diet. Histodine is an essential amino acid for infants, but its essentiality for adults has not been conclusively demonstrated (NRC 1980).

Although protein is available from plants, the quantity and quality from animals is greater. Many typical plant foods, for example, cannot provide critical amino acids. Some of these essential amino acids are required in order to build our own proteins. These amino acids can be provided by plants only if legumes, such as soybeans and peas, are added to the diet, but this must be done on a daily basis. People do not store amino acids very well, and excesses are quickly converted to lipids or used for energy on a daily basis. (The oxidation within the body of 1 gram of protein in pure form yields about 4 kilocalories.) The deficiency of one or more of such amino acids from the daily diet can lead to such serious protein deficiency diseases as kwashiorkor. This malady afflicts 25 to 50 percent of the children in developing regions (especially Africa and the Near East) because the people cannot afford animal protein or are not getting a balanced diet.

Vitamins are essential organic compounds that the human body cannot synthesize for itself. They are required in small quantities in the diet or can be absorbed by the intestine after they have been produced by the bacteria living there. The vitamins are required by enzymes in order to function (as cofactors of enzymes). Thus, the absence of a particular vitamin may result in various diseases such as beriberi (lack of vitamin B-1), scurvy (lack of vitamin C), and rickets (lack of vitamin D).

Many inorganic nutrients or minerals are as essential for good health as the organic nutrients, and they too are required in the diet. Minerals needed include water and ions of sodium, chloride, calcium, phosphates, iron, and iodine. Most of them are obtained by humans directly from plants or from animals that have eaten the plants, or from drinking water (some minerals are dissolved in water). Various symptoms develop when minerals are lacking in the diet such as stunted growth (lack of calcium), iron-deficiency anemia, and goiter (lack of iodine).

World Food Production. The ecologist K. E. F. Watt (1968) has postulated the ultimate capacity of food production of the earth using the following ecological criteria:

1. The quantity of light energy that falls on a unit area of land per year.
2. The number of units of land on the surface of the earth suitable for cultivating crops, and the number of units of water in all streams, rivers, lakes, and oceans.

3. The percentage of light energy that is incorporated as chemical energy in each crop.

4. The number of energy units required to feed an average individual for one year.

Knowledge of (1) and (3) above allows computation of the amount of energy captured on a unit of the earth's surface in a year for each type of crop. This result, when combined with (2), yields the total amount of energy that could be captured for each crop. Division of the quantity by (4) yields the total human population that could be supported by each type of combination thereof. Projection indicates that the earth's human food resources can sustain about ten billion people (table 3-1). This estimate is the "maximum that an intensively managed world might hope to support with some degree of comfort and individual choice" (National Academy of Sciences 1969).

According to recent reports, the quantity of land utilized for crops on earch is increasing but at slower rates than before (CEQ 1981). Millions of hectares are brought under cultivation for the first time each year, but nearly as much is taken out of cultivation to be urbanized, returned to pasture or forest, or abandoned. With the earth's population growing exponentially and increases in total arable land minimal, arable land per individual is declining.

As figure 3-1 indicates, world agricultural production has grown linearly despite the relatively small increases in arable land. In developing countries, however, production per capita is not increasing as much as in developed countries. It seems that developing nations may be inadvertently adhering to the Malthusian doctrine. The eighteenth century clergyman Thomas Malthus observed that population increase among the poor was exponential (1, 2, 4, 16, . . .) whereas food-growing increase was by an arithmetical progression (1, 2, 3, 4, etc.). The result, he predicted, was inevitable — periodic famines, plagues, and wars.

Today the great Irish Famine of the 1840s is largely forgotten. This famine is reputed to have reduced the population of Ireland by 50 percent, and over one million people died of starvation or related diseases (Schneider 1983). Nevertheless, it stands as an example of the dangers of unrestricted population growth and overdependence

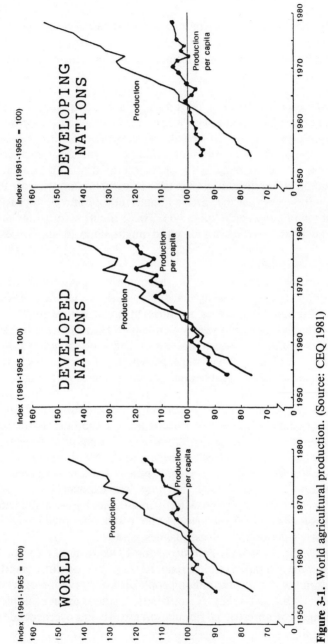

Figure 3-1. World agricultural production. (Source: CEQ 1981)

on a single source of food (e.g., potatoes). Is there a similar danger today? A considerable part of our planet relies on cereals, principally rice, maize, and wheat. Only the United States, Canada, and Australia consistently produce a surplus of these staples for export. What if a blight were to strike the grain fields of these three nations? If losses from climate, mildew, rust, smut, etc., were sufficiently great to prohibit the export of cereals to other countries, famines of much greater proportions than that suffered in Ireland would certainly occur. Even without future blights, these three countries cannot indefinitely serve as bread basket to the world. Note, for example, that the drought-damaged U.S. grain harvest in 1988 fell below consumption perhaps for the first time ever (U.S.D.A. 1988).

The world's supply of arable land, and therefore of potential food, will soon be inadequate to support future generations. The influence of technology in the form of better fertilizers, pesticides, and high-yield seed grains, as well as improved mechanization and management techniques (food from the seas, etc.) can increase present productivity. Yet current trends will not continue. All available evidence indicates that "the nineties will be unlike any decade that the world's farmers have ever faced" (Brown et al. 1989). There will be diminishing return in the use of fertilizer; many pest species are becoming resistant to pesticides; high-yield plants are vulnerable to diseases; and the problem of pollution by fertilizers and pesticides threatens supplies constantly. Also, the availability of water may become the single most important constraint to increasing yields in developing countries. Moreover, "most of the elements that now contribute to higher yields — fertilizer, pesticides, power for irrigation, and fuel for machinery — depend heavily on oil or gas" (CEQ 1982). In addition, as exportable petroleum and other fossil fuels dwindle, food-exporting nations will be tempted to convert their exportable surpluses of grain, sugar, and other foodstuffs into biofuels (Brown 1982).

Lester R. Brown, Project Director of the Worldwatch Institute, points out that the future of food production is being shaped increasingly by climate and resource constraints:

In sum, overall food security is being threatened by two trends. One is the loss of momentum in the growth of output, a loss that is particularly noticeable in major developing countries, such as China, India, Indonesia, and Mexico. The second trend is the warming of the planet. The areas that are likely to experience higher temperatures and lower rainfall include some of the earth's key food-producing regions, such as mid-continental North America. The world's farmers—already struggling to keep up with the record year-to-year growth in population—are facing the nineties with a great deal of uncertainty about how quickly the warming will progress and how it will affect their production (Brown 1989).

Land and Carrying Capacity

In the past, with the open country and the vast amount of land, no one thought of land as a finite resource. However, as the world enters into the twenty-first century, recognition of serious problems suggests questions about the carrying capacity of the land to now support the world's population. One of these problems is desertification: the process of converting arid and semi-arid lands into veritable deserts. At the present rate of conversion, desertification could significantly reduce high agricultural productivity in both irrigated and nonirrigated lands (Verstraete 1986, Postel 1989). It could also seriously lessen the carrying capacity of the arid lands to support human settlements and biota.

A second concern is the worldwide trend of converting natural forest biomes to agricultural, urban, water resource, transportation, and other developments. If this trend continues, it could seriously reduce the number and size of natural forest biomes (tropical, deciduous, and coniferous forests), which will thereby eliminate the habitat that supports most of the growth of green plants which provide food for the fauna.

As mentioned in chapter 2, the greatest threat to biota is now the decline and modification of available habitats. The problems of widespread loss of natural productive land via desertification and deforestation are reviewed in this section.

Desertification. Desertification is the spreading of desert-like conditions into areas that are more productive. It is characterized by the lowering of water tables, shortage of surface water, salinization of existing water supplies, and wind and water erosion. The most significant direct cause of desertification is overusing land in arid and semi-arid areas, including overgrazing, overdrafting (mining) of groundwater for agriculture, poorly managed irrigation, use of soils with poor drainage, and choice of crops and agricultural practices that neglect soil conservation. Other factors leading to desertification are industrial development and rapid urban growth (Postel 1989).

In her review of the desertification, Sandra Postel (1989) used the word land degradation interchangeably with desertification. She described the desertification scenario on rangelands:

Degradation on rangelands mainly takes the form of a deterioration in the quality and, eventually, the quantity of vegetation as a result of overgrazing. As the size of livestock herds surpasses the carrying capacity of perennial grasses on the range, less palatable annual grasses and shrubs move in. If overgrazing and trampling continue, plant cover of all types begins to diminish, leaving the land exposed to the ravages of wind and water. In the several stages, the soil forms a crust as animal hooves trample nearly bare ground, and erosion accelerates. The formation of large gullies or sand dunes signals that desertification can claim another victory.

Overall, desertification of the earth's land surface is increasing at an estimated rate of 50,000 square kilometers (19,305 square miles) per year. If present trends continue 35 percent of the world's land surface will be lost by the end of this century (U.N.E.P. 1984). The desertification problem is not of modern vintage, however. Three hundred years before the birth of Christ, once-fertile lands along the coast of Turkey became deserts principally because of overuse by people.

Today, with world population growing rapidly and minimal increases in arable land, arable land per capita is declining. This rapid loss of productive lands can have severe consequences for society and the environment. The arid western parts of the United States, for instance, are rich and productive; however, they are also undergoing desertification which seriously theatens their long-term productivity (Pimentel et al. 1987). In less developed regions of the world, desertification causes great human misery. Desertification in regions such as Africa and Western Rajasthan (India) causes malnutrition or undernourishment for millions of people. In the past fifty years in Africa alone, 650,000 square kilometers of land bordering the Sahara, once suitable for agriculture and grazing, have become barren desert. The Sahelian drought (in the southern rim of the Sahbara) of 1968-1974 led to the deaths of thousands of people. In addition, tens of thousands of animals starved and the Sahara expanded southward by five degrees latitude. In the past few years, history seems to be repeating itself. Another severe drought has caused the deaths of thousands of people, decimated livestock and wildlife, and has led to further loss of productive land to desert.

Some scientific-technical remedies to decrease desertification problems include improved management practices (e.g., replanting or reduced grazing) and also certain techniques which have been used to keep the water closer to upper portions of drainage valleys (watersheds) in order to restore previous plants and hydrology (U.S.D.A. 1988, Brown 1988). However, if the degradation has progressed to the stage of erosion, when soil and nutrients are irreversibly lost, those areas may never be reclaimed at a reasonable cost.

In less developed regions the scientific-technical remedies are not enough (Hanks 1988, Postel 1989). In these regions underlying the problems of desertification are fundamental socioeconomic problems of unequitable land tenure, lack of employment, and rapidly growing human populations. The rural poor must gather firewood because they cannot afford coal, gas, or kerosene. Furthermore, they have no choice but to overharvest arid and semi-arid areas for fuel and food. This is due to scarce land availability and scarcity of money which is essential in providing for their families. The value of protecting these vulnerable areas, although they will not be affected for a decade or two, appears understandably small to these peasants compared with the value of feeding their families today. Conservation is a concern for only those few altruistic individuals that are not starving. As Don Paarlberg (1988), quoting from an old Byzantine proverb, states, "He who has bread has many problems; he who lacks bread has only one problem." Thus, it seems that the principal impetus for protecting arid and semi-arid areas in poor regions must come from international aid agencies and from developed nations with scientific and economic aid programs.

Deforestation. The major terrestrial natural ecosystems are tropical rain forests, deciduous forests, coniferous forests, grasslands, tundra, chapparals or scrubs, and deserts. These ecosystems all suffer similar problems of conversion to other uses and/or damage due to pollution. However, forests are under especially intense pressure. Goudie (1986) considers the "deliberate removal of forest" as "one of the most longlasting and significant ways in which humans have modified the environment, whether achieved by fire or by cutting." According to the U.N. Food and Agricultural Organization 1984 assessment, 11.3 million hectares (one hectare is equal to 2.471 acres) of tropical rain forest are lost each year through the combined activity of deforestation, fuelwood gathering, and cattle grazing. The most damaging agricultural activity is slash-and-burn agriculture. It involves cutting down a small patch of forest, burning the native vegetation, using the ash as fertilizer, and raising crops for a year or two in the ash-enriched soil. When the nutrients in the ash are depleted, the farmer and his family move on, leaving the recently used cropland fallow until wild plants have grown again. Revegetation by wild plants restores soil nutrients decades or centuries later (depending on the nature of forest being cut). Shifting agriculture can be practiced indefinitely where soils and terrain are favorable and population density is low. Today, however, shifting cultivators in many tropical forests are shortening fallow periods and using the land too intensively. The primary reasons for increased agricultural use are the loss of forest availability to traditional cultivators because of competing demands such as large plantations and urbanization, and, because of the lack of land or jobs elsewhere, the moving into forests by peasants.

Deforestation also results when people routinely destroy trees and clear whole forests in order to obtain firewood, raise cattle, or export timber. Because of the relatively high cost of coal, kerosene, gas, and other forms of energy, firewood is usually the most important fuel in developing countries. In some of these developing countries, fuel wood provides more than 90 percent of the energy consumed. Timber harvesting (especially in tropical Asia) and deforestation for cattle raising (especially in Latin America) are steadily gaining value.

It has been estimated that primeval forests covered over 50 percent of the land area of the world. Today, nearly 25 percent of the world's land area (about 3,333 million hectares or 8.3 billion acres) is covered by vegetation that in broad terms can be classified as forest. In some developing countries, rapidly growing populations coupled with ever-increasing demands for firewood and agricultural use are eroding their forestry bases. Worldwide, there is a net loss between 1 and 2 percent of the world's forest each year. (This is about 10-20 million hectares or 24-48 million acres per year!) The disappearance rates of natural forests are far higher in some countries. Haiti was once covered with majestic forests. Today, the poor woodlands and shrub lands that remain are constantly devastated and kept from regeneration; consequently, the forests in Haiti are virtually nonexistent. Honduras has lost 35 percent of its forest over the past twenty years; Costa Rica 33 percent in about ten years; the Ivory Coast 33 percent in only eight years; and Thailand 25 percent of its forest cover in ten years. Furthermore, the biological and commercial quality of many forest areas left partially standing is deteriorating.

The rate of deforestation may have significant environmental impacts on man, as well as the entire ecosphere. In their natural state, forests regulate water flow, prevent soil erosion, and may have a major influence on regional and world climate. There is a serious concern among some scientists that extensive loss of forest cover might magnify the increasing concentrations of carbon dioxide in the global atmosphere and thus, eventually will contribute to unprecedented human-caused changes in world climate (discussed in more detail in chapter 2).

In addition, natural forest provides recreation and unique scenic beauty while also serving as the basis for natural communities that provide life support to organisms (including people). One valuable by-product of plant photosynthetic activity is oxygen, which is essential to man's existence. Forests also remove pollutants and odors from the atmosphere. The wilderness is highly effective in metabolizing the many toxic substances with which we pollute the atmosphere, and in return give us a purified condition. The atmosphere concentration of pollutants over the forest, such as particulates and sulfur dioxide, are measurably below that of adjacent areas. Food, fiber, fodder, fuel, building materials, industrial chemicals, and medicinal drugs are beneficial resources produced by natural ecosystems which, with wise management, may be renewable resources. The ecological and economic importance of maintaining biological diversity is described in chapter 2.

At this time, deforestation could be reversed in many areas by tree planting programs, removing land from agriculture use, and employing alternative nonfirewood sources of energy (such as geothermal, gravitational, wind, and solar). Similar to remedies for desertification in poor areas, the combination of economic poverty, land scarcity, and expanding human population has prevented reforestation programs from keeping up with forest losses. Consequently, remedies in poor countries require an integrated policy of scientific expertise and socioeconomic changes.

A report of the U.N. Environmental Programme provided some ideas as to how a number of simple changes in behavior and techniques utilized in burning fuel wood could significantly conserve energy. Generally, only 6 percent of the heat value is captured in the usual wood-cooking stove in Indonesia. However, when the stove is designed with a partial sunk-in-pot receptacle it conserves 20 percent of the energy; a newly designed pot saves another 30 percent; and drying the wood for several weeks before burning saves another 10 percent. The report suggests that if the overall efficiency improvements averaged 50 percent, a 10-hectare (25-acre) firewood plantation cultivated with fast-growing plants and carefully managed for sustainable yield could supply the fuel needs of an Indian village of 1,000 families indefinitely. The problem again is largely socioeconomic since most poor people don't have the money to buy newly designed stoves and pots. The few that could afford these changes have to be convinced that these items, in the long run, cost less than the firewood saved.

In 1987, the World Bank published a report evaluating the population sustaining capacity of seven West African countries (Gorse and Steeds 1987). The study focused on gauging the carrying capacity of these countries as delineated by fuel wood and food supplies. It was concluded that in the zones where rainfall is lowest, sustainable agricultural and fuel wood yields are well beyhond that for which the zones could sustainably provide. Surprisingly, fuel wood emerged as the limiting factor.

According to the report, the actual population of all seven countries in 1980 was thirty-one million. This number already exceeded the twenty-one million who could be sustainably supplied with wood resources. Unfortunately, with a projected population by the end of this century of fifty-five million, the total degradation of the land is bound to increase (Population Reference Bureau 1988).

Limits of Ecological Criteria

A fundamental problem with ecological criteria is that scientific uncertainty is pervasive. Ironically, though determining carrying capacity is one of the most crucial issues facing mankind, it is also one of the most difficult to resolve. Scientists are ever sure about what constitutes carrying capacity, about the resource requirements of a sustainable society, or about the assimilative capacity of the ecosphere. Consequently, models and predictions may not be entirely accurate because they are based on today's incomplete scientific information. Indeed, prompted by scientific uncertainty, Schneider (1983) points out that "there is no wholly 'scientific' method to assess reliably what the future carrying capacity might be since it depends on uncertainties such as future technological breakthroughs, the long-term effects of toxic wastes, the causes of birth-rate reductions, long-term climatic variations, etc." (In chapter 5, we will examine and illustrate the problem of scientific uncertainty in greater detail.)

Another fundamental problem with ecological criteria is that "one cannot assume, as some animal ecologists seem to imply, that the optimum for the human species is analogous to the optimum for an animal species" (Spengler 1971). Bergstrom (1973), for example, has made the simplistic suggestion that a "country is overpopulated when it has too many people to be supplied by the inadequate basic resources of its territory." By that definition, the United States and most developed countries are overpopulated.

Further, ecological criteria are inflexible and insensitive to human values, aesthetics, and ingenuity (Santos 1974). This point has been elaborated recently by Julian Simon (1984), who puts an incredible amount of faith in man's foresight and the rate of technological advance. To illustrate, if fossil fuels are depleted, he believes new substitutes will be found. According to Simon, the free play of market forces—including natural ones—will promote equilibrium. In his words:

Nor do we say that a better future happens automatically. It will happen because people—as individuals, as enterprises working for profit, as voluntary group, as governmental agencies—will address problems and will probably overcome, as they have throughout history. (Simon 1984)

Obviously, any model that realistically approximates the real world must incorporate the complex processes of natural ecosystems with socioeconomic systems. As Catton, Jr. (1987) recognized, "unless ecologists take the facts of human culture appropriately into account, they, just as truly, are unrealistic." In economy, the terms "bionomics" (Georgescu-Roegen 1977, Clark 1981) and "holoeconomics" (Grunchy 1947) have been coined to denote the discipline that would study the role of ecological systems as well as human-made physical systems in maintaining the overall economy. Apparently, the key issue for the future involves coupling the economy and the ecosphere. Eugene O. Odum, professor of ecology, captures the essence of this coupling in stating: "When both the 'study of the household' (Ecology) and the 'management of the household' (Economics) can be merged, and when Ethics can be extended to include the environment as well as human values, then we can indeed be optimistic about the future of humankind" (Odum 1983).

The next chapter presents a synergistic model that integrates the different ecological and socioeconomic systems in such a way as to secure maximum unity in determining carrying capacity.

4 World Order and Environmental Stability

This chapter will discuss the ecological *stare decisis* doctrine and natural law as sources of principles to guide mankind in determining what ought to be the optimum carrying capacity for a given nation and for the world as a whole. It will propose a world society in which the medium of exchange among nations would be based on an ecological currency and would be politically implemented through a World Environmental Authority—proposed seventh organ of the United Nations. The goal is to check the current inequitable squandering of the planet's resources, to stop the despoiling of its life-support systems, and to provide a global vision and strategy for solving the problems.

Louis Rene Beres (1975) distinguished four phases of what he called "world order design," which may be useful in the planning of alternative future worlds: value, hypothesis, model, and recommendation. Taken together these four stages provide a conceptual framework needed to develop alternative world patterns that are presumed better than those we have today or those we are headed toward in the future.

ENVIRONMENTAL VALUES AND PARADIGMS

The first step in the study of world order design is the value phase. This subjective and prescriptive phase provides a sense of direction for the study process and serves as criteria for evaluating the desirability of the world of the present. This chapter will propose an alternative world future in which ecological balance is maintained and concomitant effects of undernourishment and/or malnourishment are reduced. These goals are self-evidently good, or at least intuitively attractive. As the founder of modern ecological ethics, Aldo Leopold (1949), wrote:

The "key-log" which must be moved to release the evolutionary process for an ethic is simply this: quit thinking about decent land-use as solely an economic problem. Examine each question in terms of what is ethically right, as well as what is economically expedient. A thing is right when it tends to preserve the integrity, stability, and beauty of the biotic community. It is wrong when it tends otherwise.

If agreement with the above ethical paradigm cannot be reached, then world-order reasoning can proceed no further. (Thomas Kuhn, in 1962, used the word paradigm to describe the model of reality that is internalized when individuals become socialized as members of scientific communities. The model insulates scientists from problems and anomalies that may be important but are not amenable to solution within the framework provided by the paradigm.) Unfortunately, the value phase is purely subjective, not scientific. In the words of Rolston (1986), "Environmental science describes what is the case. An ethic prescribes what ought to be." This constraint places a limit on the use of scientific knowledge in evaluating the desirability of world design. This is because science cannot be used to determine if a world order design is good or bad, moral or immoral, ugly or beautiful. For example, to a male gorilla, a human female (whether Jaclyn Smith, Raquel Welch or Marilyn Monroe) is a horrible creature. Clearly, "beauty is in the eye of the beholder." When Carlo L. Lastrucci (1967) wrote "The Scientific Approach—Basic Principles of the Scientific Method," he recognized this constraint. He stated: "The attitude of the scientist toward normative (i.e., value-endowed, expressing a desirable norm or standard) phenomena is [to refrain] from inferring or implying that any phenomenon is 'good' or 'bad' per se. Whether or not absolute objectivity is ever possible for human beings to achieve is a question best left to philosophers."

As already noted, scientific principles are descriptive, not prescriptive. They do not say how things should be; they say how things are and probably will be. This objective test precludes the acceptance of subjective knowledge. A scientist may testify as to what was, is and will be. But he is going beyond science when he testifies what should be. What should be is a value judgment. For example, an ecologist testifying about lake succession can cite the evidence demonstrating that phosphates in surface waters may accelerate succession. Whether succession itself is good or bad depends. Frogs, some insects, and many swamp trees would find the swamp habitable, whereas some fish, like trout, and the person who owned the luxury hotel on the lake shore would suffer damage.

In numerous legal cases, the plaintiffs try to find out whether there is any scientific evidence of damages. Damages to ecosystems are value judgments. All scientists can do is to state with a degree of statistical probability that, for example, a continued drilling or pumping of oil can have environmental consequences (seismic hazards, oil spills, etc.). The damage that results turns into a value judgment. This limitation is quite unfortunate since most environmental lawsuits involve the measurement of natural beauty and the balancing of it against purely economic-legal-social-political considerations. Bacow and Wheeler (1984) recognized this problem when they said:

Most environmental disputes arise because people have different views over what constitutes good policy for the environment. A utility may propose to build a power-generating dam, but farmers and conservationists fight it because of its effect on irrigation and wildlife downstream. The government may license a new regional landfill that is opposed by neighboring residents who fear the noise it will generate. By adopting a new regulation that requires municipalities to improve their wastewater treatment facilities significantly, the Environmental Protection Agency (EPA) may unwittingly invite opposition both from the affected cities who claim the regulations are needlessly stringent and costly and from environmental groups who argue that the regulations are not strict enough to protect water quality.

The fact that ecology cannot be a source for all environmental values does not mean that it cannot be considered a foundation of some values, or indeed, of environmental value per se. Mark Sagoff (1985), for example, suggests that "The law calls upon ecology for guidance not just to make environments more productive, but also to protect ecosystems for their intrinsic natural qualities." He goes on to emphasize this by saying that "Ecologists then serve policy makers not only by helping achieve given objectives, to increase the profitability of farms and fisheries, for example, but also by helping them decide what their objectives should be, i.e., what to preserve and why."

To illustrate the direct influence of science on public policy, consider the history of laws regulating health. After Jonas E. Salk developed vaccine for poliomyelitis, some governments decreed that all children entering or already in school may be vaccinated. Once ecologists perceived that estuaries are generally productive and marine life concentrates there because of high nutrient conditions, these data formed the foundation for most laws aimed at protection of these coastal wetlands (Cooper 1982).

Consequently, it seems that once scientists are capable of determining for a fact that the maintenance of the integrity of the ecosphere is something that is healthful for society, governments may begin to enact laws curtailing the used of nonrenewable resources and emissions of transboundary pollutants. The rationale again will be the long-accepted vital social objective of the state's compelling interest of protecting the health, safety, and welfare of the public.

Moreover, according to U.S. Federal Rules of Evidence, once a particular scientific principle has become sufficiently well-established (i.e., normally accepted among scientists), the legal system no longer needs proof (expert testimony) of the underlying basis of the theory. In other words, theoretical justification of a scientific theory is less important than whether the theory is widely accepted by scientists. Let us examine in some detail the role of law in the proposed synergistic model.

HYPOTHESIS AND LEGAL PRESUMPTIONS

The second stage of the world-order model is referred to as the hypothesis phase. The task of this phase is to link values with a variety of factors that will more fully realize the goals set forth in the first stage. These presumed linkages or connections are hypotheses. They propose speculations about the factors that might be important in supporting the values.

This chapter will propose an alternative future world compatible with the structure and function of natural systems in which human values based on scientific principles are presumed supreme over human laws based on economics, sociology, and other concepts of human civilization. The presumption operates, until rebutted, to shift the burden of production of evidence that ecological laws are not supreme to mankind. The presumption is established to serve as an ecological policy device. It operates to favor the contention that "nature knows best," and correspondingly handicaps the laws based on socioeconomic criteria. The presumption is destroyed by evidence that "nature does not know best." The trier of fact (administrative agency like an environmental protection agency, judge or jury) must be persuaded of the truth of the disputed fact, whether nature knows best, by evidence beyond a reasonable doubt (probability less than 0.05). When a probability or significance level of 0.05 or less is obtained, this implies that an event normally can be expected to occur only 5 percent of the time or less often by chance.

For example, one of the most important human parasites is the protozoan *Plasmodium vivax,* the causative agent of one type of malaria. There is a presumption that this protozoan has natural rights which grow out of the nature of the universe. Among the rights endowed by nature or ecological doctrine is the privilege not to be destroyed as a species by human action (we are assuming that humankind could someday destroy this species). The presumption that *Plasmodium* has a fundamental right not to become extinct can be overcome by proof beyond a reasonable doubt ($P < 0.05$) that this species causes the death of two to four million people each year. Thereby, a compelling human interest exists that supersedes the right of *Plasmodium.* On the other hand, there is a presumption that other natural objects (natural ecosystems, benign species, etc.) have the privilege of maintaining their natural integrity or structure and function in the ecosphere. Humankind does not have an absolute right to destroy these natural objects. The legal rights of natural objects were summarized in chapter 2.

This ecological *stare decisis* doctrine would maximize the values of maintaining ecological stability as well as avoid the human suffering that results from undernourishment and malnourishment. The doctrine involves the assumption that, like all other species, Homo sapiens is not exempt from ecological constraints: humankind is an organism and as such can be viewed in ecological context. For example, our society can choose to use fossil fuels, but it cannot choose that they will not be depleted or that waste heat will not be generated due to the second law of thermodynamics. The laws of nature are mindless and mechanistic, amoral and precise. Many authors have seen a link between the science of ecology and human values. The ecologist Odum (1971), for example, wrote: "We can also present strong scientific and technological reasons for the proposition that such a major extension of the general theory of ethics is now necessary for human survival." The international lawyer Falk (1971) commented on the need for a new synthesis of ecology and ethics:

Such a posture of concern and proposition makes of human ecology a kind of ethics of survival. It is a science that relies on careful procedures of inquiry, data collection, and detailed observation as basis of inference, explanation, and prediction. But it also involves a moral commitment to survival and to the enhancement of the natural habitat of man.

When the bioethicist Potter (1971) wrote *Bioethics, Bridge to the Future,* he discussed the synthesis of human survival and ecological ethics:

Mankind is urgently in need of new wisdom that will provide the "knowledge of how to use knowledge" for man's survival and for improvement in the quality of life. This concept of wisdom as a guide for action—the knowledge of how to use knowledge for the social good—might be called Science of Survival, surely the prerequisite to improvement in the quality of life.

Similarly, the ecological *stare decisis* doctrine bridges ecological ethics and administrative expediency. The doctrine combines ecology and legal knowledge and forges an administrative framework that will be able to set a system of legal priorities.

The ratification of this doctrine may be fostered by the ecological paradigm that is emerging in the context of international politics. According to Sprout and Sprout (1971) the paradigm involves a new ecological way of understanding things that stresses international politics as a "system of relationships among independent, earth-related communities that share with one another an increasingly crowded planet that offers finite and exhaustible quantities of basic essentials of human well-being and existence." William Ophuls (1977), however, takes the opposite view. He argues for a strong world government, but contends at the same time that "the very ecological scarcity that makes a world government even more necessary has also made it much more difficult of achievement."

It is tempting to compare the kinship between the ecological *stare decisis* doctrine and natural law. Beginning in the sixteenth century, scholars first differentiated natural law (*jus naturale*) from positive law or the law practiced by nations (*jus gentium*). Natural law was considered the justice thought to be the right of all humankind. The ultimate source of natural law is to be found in nature rather than in the rules of society (Couloumbis and Wolfe 1982). Positive law accepts the notion that nations actually have provided more relevant norms for the conduct of international relations. Montesquieu, in his classic book "L'Esprit de Lois" (1748), argued that natural laws predated society and were superior to those of religion and of the state.

A third major school of thought concerning the sources of international law is the eclectic (or Grotian), which holds a middle ground. The eclectics argue that both "Natural law and positive rules were sources of law" (Grieves 1977).

One of the central assumptions of natural law and ecological *stare decisis* doctrine is that there exist intelligible and immutable principles that transcend the will and consent of all ideologies. It is an interesting thought that natural law and ecological doctrine are the only realms of justice in which universal consensus is perhaps approachable. Religion, philosophy, art, music, and politics are based upon faith, opinion, sentiment, and intuition. They have a tendency to divide rather than unite humanity. Only natural law and ecological doctrine are exclusively based on natural systems.

One of the basic challenges of such a notion, however, is that students of natural law and ecological doctrine must search for an acceptable method by which one can identify such things as fundamental principles and immutable laws of nature.

Three fundamental ecological principles are offered here as material and relevant to the issue of optimum carrying capacity:

1. All organisms, including humans, have a biotic potential that is exponential.

2. As the population of any species approaches the level of optimum sustainable size, or carrying capacity, environmental resistance becomes greater and greater.

3. In some organisms, the population levels off at the carrying capacity (figure 1-4); in others, the population overshoots the carrying capacity and then the population falls dramatically (figure 1-5).

These three basic laws of nature were described in chapter 1. In chapters 1 and 2, we were concerned with the ecological characteristics of human populations and systems. In chapter 3, we drew attention to the various criteria that can be employed in calculating the carrying capacity. With this background, we can now turn to the question of how the various human and ecological systems can be incorporated into one model. In the following two sections, we shall also draw attention to the ways in which administrative agencies make environmental laws.

THE SYNERGISTIC MODEL

Dr. Peccei, founder and guiding scholar of the Club of Rome, defined mental models as "models that the human brain employs to judge situations, prospects, and actions" (Peccei 1982). A model of world order is determined by a particular hypothesis (Beres 1975). The model is created for the purpose of examining the hypothesis. As expressed in the previous section, this chapter postulates that environmental stability will be enhanced and the effects of overpopulation/pollution decreased to the extent that natural laws are supreme; therefore, the most appropriate model would be one in which scientific criteria determine optimum carrying capacity. Several theoretical efforts have been made to calculate the optimum human population sustained by the world's ecological resources (see table 3-1).

Let us suppose that public policy determines that food should set the maximum carrying capacity for any given nation and for the world as a whole. (It is noteworthy that one of the original "environmentalists," Thomas Malthus, argued that the ultimate check of human population was famine.) Since no supreme world government exists, and individual nations remain sovereign within their boundaries, it is important to consider the maximum population each nation can sustain. These figures can, in theory, be determined by applying to each individual nation of Watt's formula for the ultimate carrying capacity, together with calculation to determine the production of nonvegetable sources of food (e.g., cattle, fish). This formula does not, however, consider all aspects of a complicated international situation.

Three major areas not covered include trade, effects of pollution, and lack of scientific knowledge of life-support systems. Even more important than the above areas are the political agreements among nations. In an attempt to present a synergistic model regarding optimum carrying capacity we shall now consider suggestions regarding these areas.

The currency of trade, for example, in a synergistic model would be based upon calories. Let us assume, optimistically, that nations agree that the average person in a stable population structure requires 3,000 kilocalories per day. (The United Nations defines adequate consumption as 3,000 kcal/day for a man and 2,200 kcal/day for a woman.) For the average kilocalorie requirement per person per year we will coin the term homitroph (Latin *homo* + Greek *trophe* nourishment). Officially issued homitrophic currency —paper money backed by the homitrophic production of the nation—could serve as a medium of exchange, a measure of value, a payment for goods and services, and as a settlement of debts.

It is understood that the amount of plant production that will support a vegetarian, or near-vegetarian, human population will support only one-tenth of a population subsisting largely on meat (Deevey 1956). This is because, as explained in the next paragraphs, the transfer of energy from lower to higher food levels is extremely inefficient.

The second law of thermodynamics states that the transformation of energy is not 100 percent efficient and that as energy is changed from one form to another, some of it is lost as heat. Thus, even though energy cannot be destroyed, every transformation decreases the energy available to do work. As coal is burned to provide electrical energy in power plants, for example, only 35 percent of the energy is captured as electrical energy; the other 65 percent of the chemical energy from the coal is dissipated into the environment as hot water and hot gases. The passage of energy from vegetation to animals occurs mainly within organic compounds. Light energy from the sun is converted by photosynthetic organisms into chemical energy of organic compounds. The compounds are eventually consumed by animals or destroyed by microorganisms. (Some, however, are converted to fossil fuels.) The metabolism of glucose by organisms releases about 40 percent of the total energy as biologically useful energy. The remainder of the energy is lost as heat. (Heat energy is generally useless, but organisms can benefit from heat production in cool climates. Birds and mammals conserve "waste heat" from energy conversions inside their bodies.)

The importance of the second law of thermodynamics to human food chains is that there is less energy available to mankind at each food level of the food chain. For example, as light strikes the leaves of agricultural crops, only 1 percent of the light energy is transferred into chemical energy by chlorophyll during photosynthesis. Most light is reflected or scattered or is in the form of wavelengths that plants cannot use at all. When a livestock consumes plants, only about 10 percent of the energy from the plant is passed on to the livestock. When a person eats the livestock, again only about 10 percent of the energy in the livestock is passed on to the person. Most of the energy is lost to the consumer because organisms must metabolize, grow, reproduce, and evolve. These processes require energy.

The result of this reduction in energy is a reduction in the energy flow supportable in each food level. As alluded to already, if humans eat plants directly there will be about ten times less energy loss in the flow. In fact, many overpopulated countries (China, India, etc.) are able to subsist because their people are living primarily as herbivores, thereby saving energy by shortening the food chain. But even if humans live as herbivores and get enough calories, there is a concern with nutrition at the herbivore level. As mentioned in the previous chapter, many common plant foods cannot provide certain essential amino acids, and the deficiency of such critical amino acids from the diet can lead to such chronic protein deficiency diseases as kwashiorkor.

The picture also becomes complicated by the age distribution of each nation. Furthermore, the units of land on the surface of the earth suitable for cultivating crops are also a function of technology, weather patterns, and other factors (see chapter 3). Hardesty (1977), for example, pointed out that the "steep mountain slopes of the Peruvian Andes could not be used for farming without significant modification but were used by the ancient Peruvians after the introduction of terracing technology." Changes in climate can, of course, affect the food production of a nation. Thus, the food production of an individual nation is dynamic and requires continual re-evaluation. The task of determining homitrophs per year, the eating habits, and the age distribution of each nation, consequently, requires for each specific year and nation a vast amount of concrete data that necessarily falls ouside the scope of this book.

Based on kilocalories, a linear correlation exists between the number of homitrophs produced by a nation and the maximum number of people it can sustain (Kh). Therefore, nations that have the same homitroph production may be permitted to have the same number of people; and yet, one may be above the optimum carrying capacity (Kopt) of its territory. The United States, for example, would have a high homitrophic production and yet might exceed the Kopt of its territory if there existed a threat to the global commons resulting primarily from transboundary pollutants. To prevent this situation, we need another parameter: pollution limitation (Kp). Pollution will be considered in this case as "a destructive change in the ecosystem" or as occurring when an ecosystem's "natural feedback mechanisms cannot absorb or replace the components which our presence may add to it or remove from it" (Santos 1974). For example, river pollution can destroy the ecological ability of a river to renew its capacity to dissipate pollution, while deforestation may leave the soil exposed to mineral runoff. Pollution limitation would also place a ceiling on the maximum number of people that a nation could sustain. Unfortunately, the concept of ecosystem pollution seems certain to remain hazy and unrealistic for some time to come. Recently, however, some literature is beginning to quantify and qualify ecosystem pollution. Cairns, Jr. (1977), for example, used the term "aquatic assimilative capacity" to express the ability of an aquatic ecosystem to assimilate a substance without degrading or damaging its biological integrity. He defined ecological integrity as the maintenance of structure and function characteristics of that system.

If we expand Cairn's concept of ecological integrity to include all ecosystems within each nation, we have the concept of Kp—that is, the maintenance of ecological structure and function characteristics of each nation. It establishes in time and space what Peccei (1982) called a "modicum of internal coherence and live-and-let-live compatibility with the world ecosystem." The final determination of the actual Kp for each nation, and for the world as a whole, will have to await the completion of monographic research.

Figure 4-1 presents a hypothetical country (Nation A) that produces 100 million homitrophs per year but has a Kp limit of 80 million. Its population can safely increase to only 80 million. Nation A may opt to trade its excess homitrophs (e.g., cash crops) for manufacture products and/or natural resources (fossil fuels and minerals).

Figure 4-1 also presents another hypothetical country (Nation B) that produces 100 million homitrophs per year and has a Kp limit

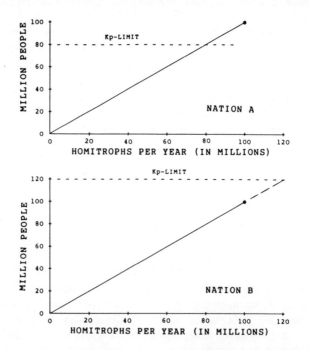

Figure 4-1. Ecological deterioration is not just a function of population, but more importantly of pollution. Nation A produces 100 million homitrophs/year but has a Kp limit of 80 million. Nation B produces 100 million homitrophs/year and has a Kp limit of 120 million. The implications are explained in the text.

of 120 million. This nation may have a low population density and/or excellent pollution control technology. Nation B can increase its population to 100 million; however, if it is able to acquire surplus homitrophs, it can safely increase its population to 120 million. In other words, if a nation were to trade foodstuffs for automobiles, that nation would be required to lower its population (by emigration and/or reduced growth rate) in an amount equal to the number of homitrophs exported by the trade. On the other hand, if a nation were to obtain a surplus of homitrophs through trade, that nation would be able to increase its population (by immigration and/or increased growth rate), provided that it did not increase beyond its Kp. Other possible alternatives include being able to reduce pollution as a result of more effective recycling plants, waste-water treatment plants, electrostatic precipitators, spray collectors, scrubbers, etc. In addition, as exportable petroleum and other fossil fuels dwindle, food-exporting nations will be tempted to convert their exportable surplus grain and other foodstuffs into biofuels (Brown 1982). In effect, there would be different levels of optimum carrying capacity depending on the degree of intellectual, moral, and personal values of each country. As a result, the setting of the optimum carrying capacity is left to the national authorities. The different levels correspond to the equity dimensions of the carrying capacity, as defined by Catton, Jr. (1987). (See chapter 3 for a review of Catton's equity dimension of carrying capacity.)

The pollution limitation factor cannot be underrated. Malnutrition and malnourishment require: (1) that population increases slow down, (2) that we increase food production, (3) that an equitable food distribution be devised, and, most importantly, (4) that the ecosphere remain stable.

Ecological stability should have two broad objectives. The first is to maintain natural processes and life-support systems on which human survival and development depend. This requires regeneration and protection of biomes, resource recovery, and controlling transboundary pollutants. The most threatened life-support systems today are agriculture, forest, arid and semi-arid lands, and coastal and fresh water biomes. The second goal is to preserve biological diversity. As discussed in chapter 2, diverse species are critical to the structure and functioning of biomes. Biological diversity also makes it possible to improve plant varieties and breeds of domestic animals, provide new foods, fibers, drugs, and countless other benefits.

It would be a Pyrrhic victory, indeed, if the acquisition of more and more food caused an ecological catastrophe. This general viewpoint has been expressed by Lester R. Brown (1971). He writes:

In considering the question of what is the optimum level as far as food supplies are concerned, it is important that we ask the right questions. The relevant question is no longer, "Can we produce enough food?" But, "What are the environmental consequences of doing so?" Unfortunately, although, from a technological standpoint, we can produce more than enough food, the ecological consequences of doing so are only beginning to appear. If we succeed in producing more than enough food, but destroy our environment in the process, what will be gained?

The scientific problem of achieving pollution control can best be dealt with by expanding basic research and development programs. As will be pointed out in the next chapter, the result of lack of knowledge regarding life-support systems can be devastating. To that end, some scientists have tried to develop computer models that bring together the interaction of global variables into one general model of world environmental conditions reaching into the next century (Meadows et al. 1972, Measarovic and Pestel 1974). Unfortunately for the present, these models are lacking in sophistication and/or are incomplete.

Scientists must aid in establishing the rules governing environmental quality. Standards must be justified and enforced. Ecological knowledge must be transmitted to the appropriate legislative bodies in order to ensure that laws are passed and enacted. In order to meet these goals and objectives, talented individuals must be attracted to environmental research. Government encouragement, including research grants and fellowships, is indicated.

RECOMMENDATION

The objective of this phase is to decide whether or not to recommend the model. Beres (1975) proposed two criteria that may be applied to evaluate the model: desirability and implementation.

Desirability

Desirability apparently depends on the initial value that prompted the inquiry. The value stated in the first phase is the maintenance of ecological stability and the concomitant reduction of undernourishment or malnourishment. According to Grieves (1977):

States support most international law for very basic reasons: (1) logic—many rules (e.g., standardized lights for steamships) simply make sense if there is to be any international order; (2) fear—there is the danger of reprisals from other states if agreed-upon rules are broken; (3) self-interest—states clearly benefit from the rules; and even (4) morals—although this is not a dominant aspect of international relations, appeals to a higher standard of community values do appear in national declarations of foreign policies.

Thus, nations may consider the ecological *stare decisis* doctrine desirable if they are convinced that international ecological problems exceed the capacity of any national government to cope effectively by itself and if their distinct interest is better served by such a model than by the dangerous chaos of the laissez-faire management of the global commons by individual nation-states. Management is especially needed for the control of transboundary pollutants emitted by the industrialized nations. One might add that there are two principal causes for the poor nutrition of people in the underdeveloped countries: their numbers are too great to be sustained by their agriculture and industry, and their agricultural methods are inefficient. Therefore, morally, it would be better for all concerned if some method were devised to prevent mass starvation. Of interest is the possibility that the acceptance of the ecological *stare decisis* doctrine may encourage the development of individual rights to diets that do not fall below fixed nutritionally safe standards and of a global program of wealth redistribution.

The statement by Maurice Strong (1987), the first head of the U.N. Environmental Programme, gets to the heart of what is at stake:

People have learned to enlarge the circles of their allegiance and their loyalty, as well as the institutions through which they are governed, from the family to the tribe to the village to the town to the city to the nation state. We are now called upon to make the next step and final step, at least on this planet, to the global level.

Implementation or Feasibility

Let us consider the present political situation and how optimum carrying capacity could be implemented. The model described in the previous section provides the ecological matrix for reaching the carrying capacity. Administratively, the model could be implemented through a World Environmental Authority (WEA)—a proposed seventh organ of the United Nations. The WEA could, with much stretching of the imagination, be equipped with administrative powers. This, of course, requires radical changes in the United Nations. Today, not even the U.N. General Assembly comes close to being a true legislative body. Its resolutions have neither legal nor substantive binding power. Furthermore, the International Court of Justice hears only those cases referred to it by consenting governments or international organizations, and no executive authority exists to enforce its decisions (Rosencranz 1983). In addition, there would be the problem of convincing many in the global community that the WEA was not a scheme to erode their sovereignty on behalf of big-power economic and political interests. The limits of international law will be examined in detail in chapter 5 dealing with the problems and prospects of achieving a stable global commons. The purpose of this section is to explore the legal rules regarding the WEA's exercise of governmental authority and the state's rights in relation to WEA authority.

Administrative Law and the WEA. In recent years, due principally to the growth and function of the government, the developed and developing nations have enacted numerous laws, especially for the purpose of protecting public health, education, housing, and other public services. The responsibility for continuous and systematic administration of these laws, as well as other legislation, have been delegated to the many public agencies created for that purpose. Consequently, an administrative agency is a body of the government charged with "administering particular legislation" (Vago 1981).

Administrative rules may be regarded as an essential feature of modern government, in which many matters are too technical or scientific, too detailed, or too subject to frequent modification, or too dependent upon unpredictable emergencies and circumstances to be incorporated in the main body of legislation (i.e., the constitution or statutes), which is less easy to change than administrative rules.

The theory behind the organization of the new World Environmental Authority is the comprehensive international environmental management of the global commons. As mentioned in chapter 2, more and more evidence has accumulated that the separate constituents of the ecosphere are really not independent at all, but are interdependent in both apparent and unpredictable ways. Air, water, pollutants, and other molecules and compounds do not stop at national boundaries. Ultimately, we cannot separate the biota, chemistry, physics from the anthroposystem, because events in one part of this ecosphere are closely linked to happenings elsewhere.

Consequently, uncoordinated efforts to control the international commons are certain to be inefficient and ineffective. It is important that we treat various environmental problems as interdependent, and too scientific, or too complex, or too subject to frequent modification, or too dependent upon unpredictable circumstances to be left to individual nations. As Contini and Sand (1972) stated:

Environmental problems characteristically require expeditious and flexible solutions, subject to current updating and amendments to meet rapidly changing situations and scientific/technological progress. In contrast, the classical procedures of multilateral treaty making, treaty acceptance and treaty amendment are notoriously slow and cumbersome.

Once we have conceded that the international environment must be governed, and that it cannot, therefore, be left to laissez-faire attitudes and cumbersome lawmaking procedures, the question whether we need a WEA to govern the global commons becomes moot.

Structure of the World Environmental Authority. Optimistically, let us assume that radical changes are made in the United Nations, and the WEA is established. The WEA should have a Director appointed by the U.N. Secretariat upon the recommendation of the General Assembly for a term of five years. Ideally eleven offices should be established within the WEA to coordinate and supervise environmental activities, each office dealing with a separate concern: pollution limitation, homitroph allotment, food production, air pollution control, water quality, land and carrying capacity, toxic substances, natural resources, planning and management, enforcement, and research and monitoring.

The head of each office and other functional administrative officers could be elected by the General Assembly for a term of nine years with the possibility of re-election. No two officers should be citizens of the same nation. A system of staggered elections of officers should be encouraged. All WEA officers should be professional scientists (with Ph.D.s or the equivalent).

Functions of the World Environmental Authority. The WEA should have a constitution with a set of statutory mechanisms that provide adequate standards for regulating the powers and procedures of the Authority and for facilitating its relationship with the rest of the international legal system. Its regulatory statutes should provide only very general guidelines, which would be interpreted and applied by the WEA. The broad goals of the WEA would include the establishment of an equilibrium among population, resources and pollution, and thus contribute to the establishment of international ecological integrity.

The maintenance of ecological integrity would require that the WEA keep nations from dumping transboundary pollutants in the environment in ways that endanger other nations. That is, as described in chapter 2, the problem of externalities or, as Garret Hardin (1968) termed it, the "Tragedy of the Commons." Noteworthy is that "as the world gets more crowded, it becomes less and less possible for any socioeconomic actor to gather only the benefits of his actions and to externalize his costs to others" (Kile 1982).

Many functions of the WEA have no counterpart in national activity. The WEA would function much like the universal adaptive system described in chapter 2. In carrying out such a function, the WEA is assumed to represent the collective will of the international community and to be acting for the common good. For this reason the WEA would be given powers not normally conferred on nations. The WEA would be authorized to require nations to surrender their freedom of action in many ways, ranging from economic control to population policies.

Like most administrative agencies, the WEA would be a microcosm of a conventional three-branch government to a considerable extent. Normally, the legislative branch will make and enact laws. Then, the executive branch will execute the laws and ensure the practical administration of judgments, decrees, and so forth. Finally, the judicial branch interprets the meaning of the constitution and laws that are enacted. All three of these activities would have their counterparts in the conduct of the WEA. The three principal formal processes of the WEA are investigation, rule-making, and adjudication. We shall consider these separately.

Investigation: The WEA generally would have broad investigative powers to subpoena records and information, including to collect, analyze, and interpret information concerning the amount of homitrophs produced, population figures, pollution limitation guidelines, etc. Article 104 of the U.N. Charter states that "The Organization shall employ in the territory of each of its members such legal capacity as may be necessary for the exercise of its function and the fulfillment of its purpose" (U.N. 1969). Therefore, *de jure* the WEA, under the provision, would be authorized to enter the territory of every nation to carry out its stated objectives.

The authority to investigate must be given to the WEA. Without information, it would not be in the position to regulate and protect the global commons. Most WEA actions in both formal and informal proceedings are conditioned by the information obtained through the WEA's prior investigations. The power to investigate is one of the functions that distinguishes WEA from international courts. This power is normally exercised in order to correctly carry out a further primary function, that of rule making, which we will discuss later.

The scope of investigative power would be delineated by the WEA constitution by requiring that (1) the investigation would have to serve a legitimate need of the WEA for information, (2) the inquiry must be authorized by statute, (3) the demand must be definite and not vague, and (4) the information must be relevant to the inquiry.

The above constraints to the investigative power would prevent the WEA from using the threat of investigation to achieve regulation in areas beyond its jurisdiction. Moreover, the WEA should have no power to enforce its own subpoenas. Rather, it should be required to request from an international court an enforcing order that, if violated, would be punishable for contempt of court. These provisions enable nations to obtain judicial review of WEA.

Rule-making: As previously mentioned, administrative agencies were created to deal with technical or emerging problems requiring supervision and flexible treatment. For example, the issuing of ten million homitrophs by the WEA or the establishment of 0.03 percent carbon dioxide at sea level as the international ambient air quality standard by the WEA are activities demanding technical training that members of the U.N. could not really provide. The U.N. General Assembly is composed of generalists. Therefore, the delegation of rule-making authority to the WEA permits legislation by scientists/specialists. "Rule-making is legislation on the administrative level" (Vago 1981).

Since the U.N. Constitution would provide only general standards to guide the WEA and the regulated nations, it would be desirable that the authority particularize the statutory standard by promulgating rules. As a semi-legislative body, WEA issues three categories of rules: procedural, interpretive, and legislative. Procedural rules are internal housekeeping rules adopted by the WEA to describe its methods of operation and list the requirements of its practice for rule-making and adjudicative hearings. Most rules would be interpretive rules which grow out of the necessity for WEA to have an announced policy. They would simply interpret a statute for the guidance of the WEA personnel and affected nations (e.g., if a rule permitted particular amounts of pollutants, a regulated nation could be fairly certain that the WEA would not proceed against it for acting in that way until the rule was changed.) Legislative or substantive rules are those made pursuant to statutory authority (express or implied); they would be, in effect, ministatutes.

The legal power to make rules would be one of the advantages of the WEA, as opposed to having only regulatory power. For example, the WEA would be able to put finely detailed rules into effect in advance of their application because of the constant attention the organization would give to the practical results of its rules and because of the quick action it could take in the event a limited change of rules was necessary.

The U.N. statute should require that the proposed rule be announced in advance and that interpreted nations should be given the right to a hearing. The problem might arise when the WEA attempts to make ex parte determination, or when the scope of a hearing is restricted by the WEA. A nation that has sufficient interest or rights at stake in a determination of WEA action should be entitled to an opportunity to know and to introduce its own evidence. Moreover, the statute should require that the WEA consider the evidence.

Any nation may petition the WEA to issue, change, or withdraw a homitrophic allotment or other rules. After the petition is filed, it goes to the Director's Office, where it is studied to see if it is complete and if the WEA has jurisdiction over the request. If it passes, it is routed to the Homitrophic Office, which collects data on food production. Based on this information, the Homitrophic Office makes a series of recommendations to the Director, who considers the recommendations and the data and decides whether the petition should be granted. If it is granted, the Homitrophic Office proposes a rule for approval by the Director.

Adjudication: The exercise of judicial power by the WEA is called adjudication; it is the equivalent of a judicial trial. Adjudication should involve an adversary contest between the WEA and an individual nation to determine whether the nation had violated an environmental regulatory statute (including legislative rules), or, in the case of pollution licensing, whether the nation was entitled under statutory standards and administrative rules to the remedy sought (e.g., increased pollution limitation, more homitrophs, etc.).

Let us consider an illustration of the adjudicatory process. A nation has acted in P manner, and the WEA has issued an unfair pollution practice complaint; the nation admits that it did P, but in prior cases the few WEA rules have not determined whether P is an unfair pollution practice; consequently, there are no adjudicative facts in dispute. This is a traditional example of rule-making through adjudication.

Adjudication by the WEA would differ from litigation because the authority itself would have to consider the questions of what nations had done what, when, where, how, and with what motive or intent. Adjudication would further differ from litigation because the WEA's decision-makers would have and exercise an expertise in the area being regulated. Since WEA officers would be allowed to bring their own knowledge and experience to the decision-making process, they could ignore uncontradicted scientific evidence that did not square with their own understanding and experience.

Judicial Review by International Courts. The last major legal concern is judicial review of WEA decisions. Safeguards are necessary to ensure that rule-making power of the WEA is properly exercised. The administrative rule must not exceed the delegated authority; its provisions must conform with the objectives of the parent U.N. statute; prior consultation with interests likely to be affected should take place whenever practicable; and the regulations must not contravene relevant constitutional rules and legal standards.

As indicated in the preceding section, the decision of the WEA would be, in many ways, different from the decision of a trial court, and different standards should be necessarily developed for judicial review. The WEA constitution should allow any prescribed statutory review and any applicable form of legal action. The courts, however, should condition the review by requiring that: (1) opportunity for WEA relief be exhausted before they rule on the correctness of WEA action; (2) the international courts should resolve only issues that are presented, ripe for review, in cases where a decision must be reached; (3) courts will not decide a challenge unless the nation has "standing" in the sense that it is in a position to demonstrate a concrete stake in the outcome of the suit and a direct impairment of its own rights.

The pleadings before the international courts can be classified as either fact or law. Questions of fact ought to be reviewed by a special court which is herein called the International Environmental Court (IEC). Issues of law should be referred to the International Court of Justice (ICJ).

One of the principal arguments in favor of establishing an environmental court is based on the doubt in the expertise and ability of the ordinary judges of the ICJ to deal with scientific issues of fact. These judges are generally without scientific experience, and steeped in the individualistic tradition of law, would tend to disregard the scientific element of the issue. What is required is experience in, and comprehension of, the nature of the complex global commons. It is the thesis of this writer that the policy of developing a stable ecosphere should constitute a crucial factor in determining whether an IEC should be created. Of interest is that the complexity of scientific facts has led to moves to establish "science courts" in the United States in which scientists would be questioned on issues of fact (Whitney 1973, Bugliarelli 1978).

5   *Problems and Prospects*

Modern international concerns derive from the desire to preserve and protect the order of the ecosphere and to bring human activities into harmony with ecological ones. The need for vigorous initiatives throughout the world to cope with environmental degradation has been voiced by the international Biosphere Conference convened in 1968 in Paris. It states:

Although some of the changes in the environment have been taking place for decades or longer, they seem to have reached a threshold of criticalness, as in the case of air, soil, and water pollution in industrial countries: these problems are now producing concern and a popular demand for correction. Parallel with this concern is the realization that ways of developing and using natural resources must be changed from single purpose efforts, both public and private, with little regard for attendant consequences, to other uses or resources and wider social goals . . . (human exploitation of the earth) must give way to the recognition that the biosphere is a system all of which is widely affected by action on any part of it.

In going from the rhetoric of environmental concerns to the realities of environmental laws, the global community faces problems. There are at least two such key problems. First, as alluded to in previous chapters, scientific uncertainty permeates environmental decision-making. Second, although international law is a necessary institution of the global community, the law inherently has certain limitations. These are dysfunctions which, if they are not accounted for, make international law an unsuitable catalyst for the maintenance of ecological stability. This volume concludes with analyses of these two key problems.

SCIENTIFIC UNCERTAINTY

The fact that scientific uncertainty is inherent in determining what are the effects of pollutants on the global commons are often cited as justification for taking no action now. The pervasiveness of scientific uncertainty was pointed out by the Conservation Foundation (1983):

> Scientific ignorance or uncertainty characterizes most environmental problems. We do not know much about how to predict the transport of pollutants in air or water, we do not have good ways of measuring or predicting the amount of soil erosion from a field, we do not know the habitat requirements of many endangered species, we have only rudimentary knowledge of how heat circulates inside buildings. The gaps in scientific knowledge are most severe with respect to toxic substances. Many potentially toxic chemicals have not been tested at all, many more have not been tested adequately, and, even for those that have been subjected to the most extensive tests, we are often unsure what the test mean for human health.
>
> Our monitoring of environmental problems is even more deficient than our scientific knowledge. We have no monitoring data sufficient to describe accurately the extent or developing seriousness of any environmental problem.

As the preceding reading indicates, our present comprehension of how the biota and the physical environment would respond to a variety of pollutants is certainly less than perfect. Many aspects of forecasting responses are hazy and will likely remain unknown in the foreseeable future. According to Harwell and Harwell (1989) the sources of ecological uncertainties can be categorized as relating to:

Insufficient data and understanding about ecological systems and processes; the necessity to extrapolate from relationships established for particular ecosystems, particular perturbations, and particular conditions, to the ecosystem and chemical interest; the plethora of potential indirect effects in ecosystems, which can lead to unanticipated consequences; and inherent environmental stochasticity.

The present discussion will focus on the origin and nature of scientific uncertainties, and will identify the current and prospective limits to environmental predictability. In order to illustrate the nature and limits of science, we will first consider a particular environmental problem, for example, whether an alleged biocidic pollutant causes cancer.

The Scientific Methods and Biocidic Pollutants

It is difficult to define the scientific method since the experimenter "uses any method or means" which can be conceived (Steel and Torrie 1960). We will briefly consider two of the most common: (1) descriptive or uncontrolled methods, and (2) comparative or controlled methods.

In the scientific descriptive method the scientist simply gathers data (information) without employing any special contrivances that would affect the occurrence (Ziman 1984). The facts in question are discovered by recording them as they occur naturally, without attempting to manipulate the situation. Since no attempt is made to keep the variable constant, these types of methods are termed descriptive or uncontrolled experiments.

Let us consider a hypothetical situation concerning the development of the experimenter's knowledge of an alleged biocidic pollutant. Let us suppose that our scientist wants to know if there is a mutual or reciprocal relationship between the concentration of asbestos fiber in the air and the cancer rate. The scientist takes several samples of air from various work places and determines the cancer rate of the employees. The scientist then calculates the coefficient of correlation and obtains a value of $r = 0.7$. This value permits the experimenter to say that the higher value of X (amount of asbestos in air) is associated with a greater value of Y (cancer rate). Is this association statistically significant? What is the probability of obtaining this result by chance? If the probability is very low (less than 5 percent in most studies), the scientist can reject the null hypothesis because there is no statistical evidence against the hypothesis. The null hypothesis is reached when it is found that the observed data does not differ significantly from values which are expected by chance alone (Steel and Torrie 1960). The scientist consults a statistical table and observes that the probability of getting that particular result is less than 0.01 percent (P < 0.01). Therefore, the experimenter can reject the hypothesis that the concentration of asbestos in the air and cancer rate are not statistically related. The value of P < 0.01 permits the experimenter to conclude that the greater values of X (asbestos concentration) are statistically associated with the greater values of Y (cancer rate).

The statement, however, that there is a statistical correlation between asbestos and cancer may lead an unwary experimenter or nonscientist to draw false conclusions. A statistically significant correlation alone does not necessarily mean causation (Steel and Torrie 1960). Such statements merely describe events which have been observed frequently to form a correlation. Two variables may appear to be highly correlated when, in fact, they are not directly associated with each other, but, are both highly correlated with a third variable. It is true, we suspect that increased asbestos effects an increase in the cancer rate, but we have no absolute proof. The problem is to define which statistical correlations do actually reveal causal connections (and then which is the cause and which is the effect); which reveal indirect causal relations; which represent mutual effects of another cause yet unknown, and finally, which are merely accidental correlations with little, if any, associations.

Association can exist without cause-and-effect connection. For example, there is probably some sort of correlation between the amount of asbestos in the air and the cancer rate, but it is another matter altogether to claim (in a cause-and-effect way) that increased asbestos, as such, causes an increase in cancer rate. It is possible that certain other extraneous variables might also contribute, such as differences in the nutritional history and genetics of the employees. "Although such a positive correlation is supportive of possible causal relationship between the two [asbestos exposure and cancer of the pleura], it is by no means conclusive" (Black and Lilienfeld 1984). Whether a causative relationship actually exists is something that must be inferred by the experimenter in light of other knowledge. The causal evidence presented is always indirect when using uncontrolled methods. Facts about the object are proven on the basis of an inference that other facts are true. In the language of a lawyer, the causation evidence is circumstantial.

Recognizing the difficulties inherent in the uncontrolled or descriptive experiments, scientists also utilize the comparative methods. The fundamental objective of the comparative or controlled experiments is to eliminate all variables except those directly involved in the hypothesis (Ziman 1984). Therefore, included in the experiment are control and treatment groups. In the control group, all conditions are kept constant. Members of the treatment group are exactly like members of the control group, except that one condition is varied. Thus, if one group of people receives a given treatment and the control does not, then some measurement can be made of both groups and any significant difference between the average measurements of the two groups can be attributed to that treatment. The scientific evidence is direct, that is, it is offered to prove facts about the hypothesis as an end in itself. Except for the statistical uncertainty of chance, direct scientific evidence goes directly to a material scientific hypothesis without intervention of an inferential process.

To eliminate contribution from extraneous variables, the scientist can conduct a controlled experiment in which the only factor that is allowed to vary is the one whose effect the scientist decides to examine. However, people usually make poor experimental subjects since, unlike other species, their genetic or environmental conditions and their present behavioral and cultural variables cannot be controlled. For example, the scientist cannot take twenty individuals from the same family (thereby minimizing genetic variability) and separate them into treatment and control groups to determine who will suffer from cancer as a result of asbestos exposure.

Some epidemiologic studies employ what might be called semicomparative methods. In the prospective epidemiologic study, for example, the experimenter observes the effect of exposure to a single factor upon the incidence of disease in two otherwise identical human populations (Black and Lilienfeld 1984). The investigator identifies two populations, one composed of persons who have been exposed to the alleged toxic substance and one of persons who have not been exposed. The experimenter analyzes the incidence rates of disease in each population. If the two groups are comparable, any statistical difference in disease incidence can then be related to the toxic substance. Unfortunately, these epidemiologic studies are not true, controlled experiments, because the two groups are not in fact comparable. The two populations may differ in genetics, age, sex, and previous exposure to toxic substances.

To supplement descriptive and epidemiologic studies, experimenters often carry out comparative experiments with large numbers of other organisms, usually mice or rats. The advantages of comparative experiments with animals are the ease of matching treated and control groups and the relatively short time needed to conduct experiments. Another advantage is that many comparative experiments can be done that would otherwise be impossible with people as subjects.

By using animals, scientists have implicated many toxic substances as responsible for a substantial portion of the illnesses, deaths, and environmental hazards experienced by people. A striking example is cancer. With the apparent exception of arsenic, all hazardous materials related to cancer in people have been shown to also be carcinogens to animals (CEQ 1979). In addition, studies conducted on animals have implicated about 1,500-2,000 chemical substances that are potential human carcinogens (OSHA 1979).

The main criticisms and widespread misunderstanding of animal experiments have to do with dose level, routes of exposure, benign versus malignant tumors, and threshold levels of exposure (CEQ 1979).

Dose Level. Experimental methods in which carcinogens or cancer-causing substances are administered at doses far higher than those normally encountered by a human community have been the subject of criticism by affected industries and by the general public. A recent argument concerned saccharin. Critics contended that rats were fed the equivalent of 800 bottles of artificially sweetened soft drinks per day in an experiment that indicated the carcinogenicity of saccharin (CEQ 1979). What person, they charged, would ever drink that many bottles of soda? Critics argued that any chemical will be carcinogenic if given in a large enough quantity.

Apparently, there is a misunderstanding on the part of the general public as to the causes of cancer. The majority of chemicals tested, even at high doses, are not carcinogenic. Chemicals are either carcinogens or noncarcinogens. Large doses of cancer-causing substances are more apt to cause cancer in more humans. Large amounts of noncarcinogens, such as salts and glucose, may cause damage or make people ill in some other way, but they do not cause cancer. Therefore, regardless of the dose administered, noncarcinogens will never result in causing cancer.

One reason for employing high doses is to obtain statistically significant data with a few hundred mice or rats (CEQ 1979). This is because exposures to the animals over their lifetime of a few years can be translated to people who may be exposed for many years, perhaps a lifetime. There is proof that the probability of getting cancer increases rapidly with duration of exposure. It is also because increasing the dosage means that cancer will occur more frequently, so that it can be detected in a group of 100 mice or rats. Imagine two groups of 100 rodents per group, one control group and one treated with the suspected carcinogen. If no tumors are detected in the control group, the lowest incidence in the group of treated mice that we could consider statistically significant with 95 percent confidence (P < 0.05) is 3 percent. An incidence of 3 percent would be an enormous amount in the U.S. population—7.2 million out of 240 million people. Any chemical that caused an alarming incidence of illness/cancer would probably be rejected without reservation, even if there were great advantages to its use. Even 0.03 percent is a very high chance for a large population. Besides, to detect an incidence as low as 0.3 percent, at the 95 percent confidence level, would need experimental groups of 1,000 rodents. To detect incidence levels of 0.03 percent or lower, which are closer to the lifetime risks of the more typical human cancers, experiments of hundreds of thousands of mice would be necessary. For practical and financial reasons, this is not feasible (CEQ 1979). Thus, it is feasible, in essence, to make one rodent represent a thousand by increasing the dosage.

Another reason why high doses are required in animal studies is that small rodents and people differ biologically (NRCTR 1979). Normally, chemicals swallowed by people are distributed more slowly and tend to persist longer than those swallowed by small rodents. This is because blood circulates about twenty times faster in rodents than in people. In addition, the process by which chemicals are broken down and excreted from the body is slower in people than in small rodents. Even the number of susceptible cells is larger in people than in rodents. Tests employing 160 to 3,000 mice represent about the same number of susceptible cells (depending on the type of cell) as are contained in one human.

Nevertheless, one cannot be absolutely sure that humans are more or less susceptible to disease from a given dose of a cancer-causing substance than the experimental animals. There is simply insufficient data on how to extrapolate dose-response data across species lines (Doninger 1978).

Yet another limitation of laboratory testing is that it disregards the environmental conditions on dose-response relationships. As noted by Harwell and Harwell (1989), laboratory tests:

do not necessarily represent the toxicity of individuals in the environment, because the tests do not include all of the conditions that interact with dose-response relationships, such as the environmental conditions of temperature, moisture, nutrient availability and other stresses the individual may experience in the environment, e.g., from other chemicals or other anthropogenic activities.

Routes of Exposure. An important reason for criticizing experiments which use rodents is that the method of applying a suspected chemical to rodents may differ from the expected route of human exposure. Assessments of rodent experiments are made simpler when the rodents are exposed in the same way that people are or would be exposed. However, there are good experimental explanations for sometimes using different routes in rodents (Sontag 1977). For example, when it is important to know the exact quantity of an experimental chemical given to animals in a feeding experiment, a stomach tube may be employed. It might be particularly important to use this route if a disagreeable odor or taste of a substance prevents administration of higher doses. Skin experiments or tests are frequently employed as indicators of cancer-causing factors because they are convenient. Their data have frequently been verified by feeding experiments or more difficult inhalation experiments. When using a route of administration different from that expected in people, careful thought is given to any differences in absorption, excretion, and effects of the experimental chemicals when administered to rodents.

Routes of exposure, used for different regulatory reasons, may differ for a particular chemical or factor. For example, when a chemical is handled directly in an occupational setting, absorption or inhalation is deemed by regulatory agencies not the most relevant route, but when a food additive or drug is considered, swallowing is usually the most relevant route.

Benign versus Malignant Tumors. In most instances, when substances are found to induce tumors in test rodents, both benign and malignant tumors are detected. It is sometimes argued that only malignant tumors should be considered in evaluating the statistical significance of potential carcinogens (CEQ 1979). Unfortunately, there are cases that would demand the inclusion of benign tumors in determining the potential hazards of a given substance. Some tumors pass through phases before they develop to malignancies, and certain benign tumors are precursors to malignancies. As a result, this progression may render the former argument invalid. In addition, some factors known to be carcinogenic in people, such as radiation and certain polycyclic hydrocarbons, mainly cause benign tumors in one strain of rodents and malignant tumors in other strains. The majority of scientific data suggests the detection of all types of tumors is an indication of cancer-causing potential (Tomatis et al. 1973). Furthermore, benign tumors themselves will cause damage when they happen to press on some vital organ.

Threshold Levels of Exposure. Many companies contend that it is neither economical nor essential to entirely eliminate traces of cancer-causing chemicals from their products or work environment. They argue that every carcinogen has a safe exposure level. The present collective scientific opinion is that no method exists by which to evaluate "safe" or threshold levels or exposure to cancer-causing factors (WHO 1974, NRC 1977). The major inhibitant in evaluating whether there are threshold values is the scarcity of data on the impacts of cancer-causing factors at low exposure levels. Even where a qualitative link is visible, precise quantitative estimates of risks to human beings cannot be made reliably, particularly for low risks such as one case in 10,000 or more, for example (CEQ 1979). Most information scientists have about the dose-response of cancer-causing substances is derived from employees, from patients given drugs or radiation at fairly high doses, and from whole animal *in vitro* "screening," for the most part at high levels. These high doses do not answer the question of whether or not safe exposure levels exist. On the other hand, a recent investigation of 24,000 animals with low levels, found no safe threshold levels (NRCTR 1979).

Thus, in science, despite the most ingenious materials and methods and meticulour executions of experiments, evidence is often indirect because correlation by itself does not prove cause-and-effect, and animal studies require translation of results to humans. Furthermore, the extrapolation is not specific to any individual (Gold 1986). Mutations and cancer provide no real or demonstrative evidence of the inducing toxic substance. No such actual physical evidence is available. Consequently, specific causation involves inferences on causation extrapolated from a population-based experiment, rather than individual conclusions regarding causation in the individual case.

Moreover, the statistical significance might not be precise or conclusive. The actual results might be something like 85 percent probability. (Normally, a probability or significance level of 0.05 is acceptable in most scientific studies. This means that if an event is normally expected to occur only 5 percent of the time by chance, we are willing to accept the hypothesis that a more frequent occurrence is due to the experimental treatment and not just an accident of chance.) Also, results from experiments either support or fail to support the hypothesis. However, results that support the hypothesis do not actually prove that the hypothesis is true. A hypothesis that is not consistent with experimental data is rejected. A hypothesis that is not proven false by an experiment or series of experiments is tentatively adopted. However, it may not be accepted in the future when more data becomes available if it is not consistent with the new data. There is no unconditional truth in science, only different levels of uncertainty, and the possibility always remains that future evidence will cause a theory to be changed.

Uncertainty and Transboundary Pollutants

It is relatively easy to test biocidic pollutants on mice or rats; it is much more difficult to experiment on Homo sapiens; and it is practically impossible to test transboundary pollutants on large regions of the ecosphere. Ecological effects are thus often detected by experimenting with small, artificial microcosms or by modifying small regions of the natural environment. Much of our comprehension about the ecosphere, however, comes from studying natural descriptive experiments: comparing two similar forests which differ in only one or two variables or, for example, observing the effects of phosphates on the physical and biological characteristics of a lake. Unless human technologies happen to copy or parallel such a descriptive experiment (as dumping phosphates into a lake), it is difficult to foresee their result in advance. Scientists are left with the next best alternative, and that is to monitor the environment for any indication of something going wrong. Monitoring our global commons is becoming increasingly important, and we have considered some specific cases for concerns, such as the ozone hole, acid rain, and the greenhouse effect as in chapter 2. To illustrate the nature of uncertainty in somewhat more detail, we focus our attention here on the uncertainties regarding the greenhouse effect.

Scientists have long debated how increasing levels of carbon dioxide in the atmosphere will affect the temperature of the earth. Too much is still unresolved, including the interaction effects of past air emissions, the risks of future air emissions, the type of pollution-response curve between sources and receptors of present air emissions, and the socioeconomic factors. The latter includes the relative costs and benefits of various alternative courses of action.

As described in chapter 2, the atmosphere's carbon dioxide concentration increased by 9 percent between 1959 and the present. This was largely caused by deforestation and the burning of these vegetations and fossil fuel. If we continue to add airborne carbon dioxide at our current rate, the planet's average temperature will rise. It is uncertain by how much—perhaps between $1.5°$ and $4.5° C$. This increase is predicted by the middle of the twenty-first century. Moreover, different computer models disagree how each region of the world will be affected. Major sources of uncertainties are the roles of the atmosphere, oceans, photosynthetic organisms, and soil.

The models estimate that about 55-65 percent of the carbon dioxide emissions generated by fossil fuels are presently retained in the atmosphere. It is reasonably well-known that only the marine environments are currently a significant absorber of airborne carbon dioxide, the terrestrial ecosystems being approximately neutral in components (i.e., absorbing as much as they are generating). As the marine ecosystems warm, they will very likely absorb less airborne carbon dioxide unless photosynthesis increases proportionally. If removal of trees and other photosynthetic organisms continues, photosynthetic organisms would also represent a reduced resource for absorbing airborne carbon dioxide. These two unknowns have impact on the ability of scientists to predict rates of climate modification.

The difficulty in understanding ocean physics and cryospheric relations is also related to the scientific uncertainty that still surrounds the greenhouse question. Even though it is speculative that polar ice would melt, the sea level would rise simply because water expands when it absorbs heat. If the Arctic should melt substantially, it is expected that most of the world's low-lying areas would be flooded.

To make food, a plant must obtain carbon dioxide. It is well-established that when carbon dioxide levels are raised in chambers, such as greenhouses, most vegetation respond with increased growth. Nevertheless, the potential effect of high carbon dioxide concentrations on agriculture and other plant resources remains largely uncertain. Scientists cannot predict what the effects of rising carbon dioxide concentrations will be on vegetation in natural ecosystems. One thing is certain, however: not all plants will respond in the same way. Consequently, species competition would be influenced, resulting in changes in structure and dynamics of ecosystems.

Uncertainty and the World Environmental Authority

The key concern includes not merely what the scientific facts are, but also what should be done about the facts. The distinction between what the facts are (risk assessment) and what ought to be done about them (risk management) was most concisely explained by William Ruckelshaus, former Administrator of the U.S. Environmental Protection Agency, in a 1983 speech at the National Academy of Science:

Scientists assess a risk to find what the problems are. The process of deciding what to do about the problems is risk management. The second procedure involves a much broader array of disciplines, and is aimed toward a decision about control. Risk management assumes we have assessed the health risks of a suspect chemical. We must then factor in its benefits, the costs of the various methods available for its control, and the statutory framework for decision.

As the preceding sections described, since the exact laws of cause and effect are generally unknown, especially when transboundary pollutants are involved, scientific proof is characterized by relative certainty and uncertainty. Moreover, in science, "truth" is achieved by an inquisitorial process. In science, not much is lost by discarding a theory since all theories are contingent and the substitution of a new theory to replace the old one is considered an advance (Ziman 1984). The proposed World Environmental Authority (WEA), on the other hand, is a legal entity. It would place a greater value on risk management than on the truth. Consequently, despite the scientific uncertainty, the WEA must manage the global commons. Furthermore, a characteristic of the WEA not shared by the sciences is that the quest for knowledge and risk management are combined. Thus, the current paucity of information ought not prevent the WEA from adopting a strategic approach based on risk management.

Harwell and Harwell (1989) appropriately point out that one of the goals of science is to reduce the scientific uncertainty of risk assessment:

In essence, our capabilities to predict ecosystem responses to a particular stress constitute a translucent crystal ball, sufficiently clear so that the nature of stress responses can be discerned, sufficiently cloudy so that precise predictions are not possible, and surprises are inevitable. The task of ecological research is to enhance the transparency of the crystal ball, i.e., to improve our methodologies for risk assessment; the task of decision making is to make optimal decisions based on the existing clarity of the crystal ball, i.e., to optimize our risk management.

Again, scientific knowledge is essential to clarify the "crystal ball." It would be invaluable in the shaping of the WEA decisions. The WEA, however, sets the substantive and legal framework by which the scientific evidence is evaluated. Furthermore, at times, judgment must be used where solid scientific evidence is unavailable. In that case, one has a continuum from the scientific "data" to general "agreement" to "assumptions" (e.g., plants require carbon dioxide for photosynthesis) to "science policy" (e.g., should we include predicted deforestation in forecasting the warming of the earth) to "policy" (entire regulatory package including cost/benefit analysis).

LIMITS AND SCOPE OF INTERNATIONAL LAW

Another major constraint upon resolving the international environmental crisis derives from the limits and scope of international law. International law has many definitions, some of which are overlapping, and different scholars emphasize different combinations of definitions. International law has historically been considered as "the system of right and justice which ought to prevail between nations or sovereign states" (Vattel 1835). In contrast, a modern and more complex definition by Weston, et al. (1980 proposes that international law should be considered not as a set of rules applicable to nations, but instead as "a configurative process of authoritative and controlling decision which, through interpenetrating medley of command and enforcement structures, both internal and external to nation-states, effects value gains and losses across national and other equivalent boundaries."

For an understanding of the role of international law in the maintenance of the global commons, it is important to be at least aware of these different definitions and scope. But international law, like the sciences, possesses certain limits or dysfunctions. These limits stem in part from the law's reactive feature and horizontal nature, and we shall now consider them briefly.

Legal Systems Are Reactive

As indicated earlier, conceivably, statistical proof of the warming of the planet or other catastrophe might come after the time has passed when action could be taken to reverse the trend. Consequently, if environmental change is to be minimized, legal action may have to precede rather than follow the cataclysm. In the proceedings of the U.S. Strategy Conference on Tropical Deforestation, the biologist George Woodwell (1978b) noted this concern: "If we wait until there is absolute proof that the increase in CO_2 is causing a warming of the earth . . . it will be 20 years too late to do anything about it."

Recognizing this situation, some policymakers are beginning to sense the urgency. The foreign minister of the Soviet Union, Eduard Shevardnadze (1988) observed in an address to the U.N. General Assembly, that "all the environmental disasters of the current year have placed in the forefront the task of pooling and coordinating efforts in developing a global strategy for the rational management of the environment." The foreign minister then went on to stress the lack of time, stating that "we have too little of it, and problems are piling up faster than they are solved."

The preceding notwithstanding, international law is almost always a reactive institution responding to, rather than anticipating, changed conditions. These dysfunctions derive in part from the political nature of legal decision-makers. This point was expressed by the sociologist William R. Catton, Jr. (1987):

Although ecologists recognize there are limits to ecosystem sustainability, politicians are professionally compelled to remain deaf to suggestions that growth of human activities and elevation of consumption cannot be perpetual. The ecologists' time horizons are based on evolution or succession; politicians' horizons are seldom more than two or four years away, because they get re-elected by encouraging electorates to expect them (at least in election years) to promote economic growth.

In times of crisis the law can break down, providing an opportunity for discontinuous and sometimes often painful developments. It is no accident, for example, that the United Nations was formed after the disaster of the Second World War. This hindsight is what Kieffer (1979) has termed Fontaine's Law: "We believe no evil until the evil is done, if then." He then went on to emphasize that "The environmental crisis is an extension of humankind's failure to see itself as an integral part of the global ecosystem."

Consequently, as with all major reassignments of governmental responsibility, the proposed formation of the WEA would arise from both overwhelming perception that the laissez-faire approach to the global commons is inadequate and a basic change in international values. What these changes will be is at the moment unclear, but it is the thesis of this book they must include the environmental values and paradigms mentioned in chapter 4. Briefly, they would produce an alternative world future in which ecological balance is maintained and concomitant effects of undernourishment or malnourishment are minimized. Unfortunately, as the preceding discussion indicates, environmental changes will most likely precede changes in international law. Potter (1988) believes that, "Perhaps only hindsight can energize the conversion of a viable global bioethics into law." He proposed "a generalized sequence of events," with one leading to the next, that may occur as the "fruits of hindsight":

1. Environmental damage becomes visible. . ., raising moral indignation.
2. Knowledge of these problems evolves a new discipline—ecological bioethics.
3. Moral indignation demands preventive countermeasures.
4. Guidelines are converted into legal sanctions.

Facultative Nature of International Law

It is clear that the most serious threat to implementation of the synergistic model for carrying capacity would be a lack of cooperation among nations. However, we have reached the point at which the good of the world society must prevail over individual and national rights.

There has been movement in this holistic direction; the U.N. Conference on the Human Environment, and the various U.N. pollution conferences are some examples. Unquestionably these conferences have transferred the global environmental crisis "from the realm of intellectual discourse into the field of international relations" (Puchala 1988). However, the resultant beneficial effects of these meetings to date have been limited. First, the discussions often degenerated into arguments concerning the rights and jurisdiction of individual nations. As Grieves (1975) points out, "Questions of jurisdiction might appear to be of immediate importance to environmental problems but in fact jurisdictional questions are the most basic of all, cutting across of other issues." Even the United States, in a 1981 U.N. Conference on the Global Environment, had made clear its aversion to U.N. "meddling" in national matters. The delegates affirmed the American position and insisted that the U.N. Environmental Programme implement projects dealing with worldwide and regional environmental issues and continue its limited role as information bank and catalyst.

A second important reason why the meetings have not been totally effective stems from the fact that international disputes will arise because nations will have different views over what constitutes a good policy for the ecosphere. There are developing nations who would contend that the benefits from burning coal to them outweigh the cost of the "greenhouse effect" or "acid rain."

Third, the industrialized nations have achieved relative prosperity, often by exploiting their own and other nations' resources. Many developing nations fear that restraints on the use of the global commons will make it more difficult for them to develop to the same level as industrialized nations (Stone 1984). This led to understandable resentment and suspicion of the motives of those who were encouraging conservation. Indira Gandhi, for example, argued that "Many advanced countries of today have reached their present affluence by their domination over other races and countries, the exploitation of their own masses and own natural resources" (Caldwell 1984). This gap between most of the developing countries on one side, and the developed countries on the other, as to restrictions required to control pollution in the global commons became vividly apparent at the U.N. Conference held at Stockholm, Sweden. Due to this rift, the Stockholm Conference adopted a compromise "between the goals of economic development and environmental management" (Nanda and Moore 1983). This compromise is expressed in the Stockholm Declaration on the Human Environment:

> States have, in accordance with the Charter of the United Nations and the principles of international law, the sovereign right to exploit their own resources, and the responsibility to insure that activities within their jurisdiction or control do not cause damage to the environment of other states, or of areas beyond the limits of national jurisdiction. (U.N. 1972)

Obviously, the list of reasons why nations will not completely cooperate is incomplete. One may add to this a variety of procedural inefficiencies, administrative delays, and archaic legal technologies (Feld and Jordan 1983). Arkady Shevchenko (1985) captures the essence of the dysfunctions of international bureaucracy in stating:

The Secretariat not only has problems common to any large bureaucracy but suffers from some special headaches all its own. The big ones are conflicting loyalties, different administrative traditions, and lack of a cohesive executive comparable to that of national governmental institutions. For instance, just making yourself understood in the context of U.N. procedure was not always easy: in my department there were about 150 officers of nearly fifty different nationalities.

One may also include the fact that other issues besides the environment are more important to individual countries. For example, poorer nations have relaxed pollution standards in order to encourage industrial development, while countries such as Israel and Nationalist China have reduced standards because they are fighting for their survival as nations. Partly as a result of these and other limits of international law, a mood of diminished confidence in international law pervades current literatures.

It is a solemn thought that "international law cannot draw up solutions to resource problems or make nations cooperate" (Bilder 1980). As pointed out by Rosencranz (1983) "Nations control pollution only if and when it is in their national interest to do so, and not because of any obligation under international law to do so." Nevertheless, it is only a matter of time before many nations will reach their maximum carrying capacity, and, therefore, the jurisdiction of a supreme world government is of paramount importance.

Few nations will voluntarily surrender their authority to a supreme World Environmental Authority, and yet, such cooperation is vital. At present, the world exists in a "horizontal law" in which sovereign nations interact with each other and yet remain sovereign (Kaplan and Katzenbach 1961). Thus, research has to be done in monitoring population and pollution limitation guidelines in ways that will minimize the infringement of individual nations' sovereign powers. Already there are scientific techniques that can be used to monitor the environment. The Earth Resources Technology Satellite, for example, has transmitted photographs that can be used for detecting blighted vegetation, as well as air and water polution. Presently, members of the U.N.-sponsored Convention on Long-Range Transboundary Air Pollution are monitoring pollutants with an eye to regulating them (Mott 1988).

The interaction of present-day nations is a paradox, for one cannot have order, stability, and sovereign units within a larger sovereign ecosphere. At best, the larger system will be unstable. In discussing the theory of hierarchy, Pattee (1973) wrote:

It is a central lesson of biological evolution that increasing complexity of organization is always accompanied by new levels of hierarchy controls. The loss of these controls at any level is usually malignant for the organization under that level. Furthermore, our experience with many different types of complex systems, both natural and artificial, warns us that loss of hierarchy controls often results in sudden and catastrophic failure. Simple tools may wear out slowly and predictably, but as systems grow in size and complexity they reach a limit where a new level of hierarchy control is necessary if the system is to function reliably.

Richard Falk (1971), while writing on the international environmental problem, succinctly observed:

It should hardly occasion surprise that the sovereign state—suitable for a simpler world or more nearly autonomous units—cannot be expected to cope with the tasks of our world. The scope of modern problems clearly overwhelms the jurisdiction of many national governments; but also the nationalist way of doing things is becoming outmoded, given the circumstances of the endangered planet.

CONCLUSION

This is an unusual book that does not conform to the general pattern because it deals with an almost unprecedented problem: the world is being saturated with pollution and nonrenewable resources are being depleted to a point at which life-support systems might be in danger of losing their ability to maintain environmental stability. Argument is made for the value of ecological balance in which the concomitant effects of undernourishment and/or malnourishment are reduced. The acceptance of the ecological *stare decisis* doctrine—the supremacy of ecological laws—is one of the elements or causes that contributes to producing the proposed value.

The book asserts that the existing framework of international organizations is not adequate to affect the proposed ecological doctrine. It suggests a new world economic order, based on an ecological currency such as homitrophs or energy, that is politically implemented through an environmental authority—a proposed seventh organ of the United Nations.

The dynamic model will be described by many as very attractive, idealistic, and even utopian. The writer, nevertheless, believes that nations must and will cooperate to solve the international ecological problem, and that some form of vertical international laws will, as a result, be developed. The dynamics of the global commons will eventually lead to an effective instrument for the development (through legislation) and the enforcement (through implementation) of an optimum population/pollution for each nation and for the world as a whole.

World modelling should deal with big issues and with the most controversial subjects. Otherwise, what is its justification? If the results of modelling are limited to recommendations that are uncontroversial, it follows that the recommended actions are already being taken and no influence from world modelling is necessary. (Forrester 1982)

Glossary

Abatement. A reduction, decrease, or diminution.

Abiotic. Without life; pertains to nonliving factors of the environment.

Acid rain. Rain or other precipitation containing nitric and sulfuric acids derived from air pollutants, such as industrial and automobile emissions.

Act of God. Legally, an act occasioned by natural forces (hurricanes, earthquakes, floods, etc.) without the interference of any human agency.

Age structure. The percentage of a population according to age levels. The different age levels of population expressed in percentages; often represented graphically as an age pyramid.

Albedo. The reflectivity of a body as compared to the reflectivity of a perfect reflector of the same size, shape, orientation, and distance. Usually given as a percentage.

Amicus curiae. A bystander without legal right to appear in a legal suit but who is allowed to introduce argument, authority, or evidence to protect his interests.

Amino acids. Simple molecules that contain an amino group (NH_2) along with carbon, oxygen, and other hydrogen atoms, and form the building blocks of proteins.

Anemia. A condition characterized by a decreased oxygen-carrying capacity of the blood due to a deficiency of red blood cells or of hemoglobin.

Annulment. The act of destroying the existence of a void or voidable contract and everything appertaining thereto from its beginning.

Anthroposystem. An artificial system consisting of mankind and domesticated organisms and their nonliving environment, interacting to form a self-sustainable unit.

Arable land. Land that is able to be plowed or cultivated.

Asbestos. A general term applied to certain fibrous minerals displaying similar characteristics although differing in composition.

Asymptomatic. Showing no evidence of disease.

Atmosphere. The low-density, gaseous sphere which surrounds the earth.

Atmospheric pressure. The depressing force of the earth's atmosphere acting upon the earth's surface per unit area. Atmospheric pressure at sea level corresponds to the pressure required to lift a column of mercury 760 millimeters.

Benign tumor. A noninvasive cell growth that does not damage normal tissue.

Biocide. A factor that kills organisms directly by damaging their bodies.

Biodegradable. Capable of being chemically broken down by the action of organisms.

Biogeochemical cycle. The circular passage and transformation of matter, like carbon, from living organisms to the abiotic environment and back again.

Biomagnification. The greater concentration of a nonmetabolizable and nonexcretable substance, like DDT, in an organism's body beyond that found in the environment.

Biosphere. The part of the earth that contains living organisms.

Biota. The living species in a region.

Biotic. The living factors of the environment.

Biotic potential. The maximum potential growth rate of a population under ideal conditions.

Birth rate. The number of organisms born during a particular time period.

Burden of proof. The necessity or duty of affirmatively proving a fact or facts in dispute of the legal case in which the issue arises.

Cancer. A malignant growth of cells which invades surrounding tissues causing their death or abnormal function.

Carbohydrate. An organic compound with the general formula $(CH_{20})n$ including sugar, starch, cellulose, and glycogen.

Carcinogen. Any substance or agent capable of inducing cancerous growth.

Carrying capacity. Refers to the largest population size of a given species which an environment can support on a sustained basis utilizing the resources of that particular environment.

Cause. See direct cause and indirect cause.

Cell. The basic structural unit of organisms; a cell consists of a cytoplasm encased within a membrane.

Chlorinated hydrocarbons. Hydrocarbons in which hydrogen atoms have been replaced by chlorine atoms, also called organochlorides. These include pesticides (DDT, aldrin, chlordane, and heptachlor) and numerous organic compounds (polychlorinated biphenyls).

Closed system. A system in which there is no exchange of matter or energy with the surroundings.

Common law. Law created by judicial decisions, as distinguished from law created by the enactments of legislatures.

Constitution. The fundamental law or principles of government of a nation, state, society, or other organized body of people, embodied in written documents, or implied in institutions and customs; also, a written instrument embodying such law.

Constitutional right. A right guaranteed to a person by the constitution of a government, federal or state, and so guaranteed as to prevent legislative interference therewith. Over the years the United States Supreme Court has held that individuals have other fundamental rights, such as, right of privacy, which, although not specifically expressed in the Bill of Rights, can be construed from the penumbra of the Amendments to the U.S. Constitution.

Control. That part of a scientific experiment to which the experimental factor or variable is not applied but which is similar to the experimental group in all other respects.

Death rate. The number of organisms dying during a particular time period.

Density-dependent. Description of factors, such as competition and predadation, that control population density particularly when the population size is large.

Density-independent. Description of factors, such as hurricane and freezing weather, that control population density regardless of the population size.

Developed nations. Industrialized nations that have a strong economy and sophisticated technological development; Europe, Northern America, Russia, and Japan.

Developing nations. Nations that are not yet industrialized and have little or no technological development; Africa, Asia, and Latin America.

Direct cause. A cause that sets in motion an uninterrupted chain of events which brings about a result without the intervention of an external force.

Doubling time. The time it takes for a population to double in size.

Due process. Due process generally requires that the judicial system give a notice of an accusation against oneself, and the opportunity to be heard and to defend oneself before an impartial tribunal. Moreover, the government cannot exercise arbitrary and unreasonable powers.

Ecological niche. The structural and functional role played by a particular species in its environment determined by the physiological or behavioral actions of a given species.

Ecological succession. The orderly and sequential transition from a pioneer to a climax community.

Ecology. The study of the interaction of organisms between themselves and their abiotic environment.

Economic externality. That portion of the cost not taken into account when evaluating the cost of doing business or carrying out an activity, like the cost of environmental damage (e.g., polluted air), that results from a manufacturing process or driving a car.

Ecosystem. The community and its abiotic environment treated as an interacting functional unit.

Electromagnetic radiation. Energy propagated in the form of an advancing disturbance in electric and magnetic fields that exist in space or in the media through which the wave is passing.

Electrostatic precipitator. A device installed in smokestacks to remove soot and other particulates.

Endangered species. Species in danger of extinction and whose survival is unlikely if the causal factors continue to operate.

Energy. The capacity to do work or to cause specific change. Energy holds matter together and can become mass or can be derived from mass.

Energy pyramid. A food pyramid in which the various nutritional interactions are expressed as the amount of energy required to support each level of consumption.

Energy recovery. Refers to obtaining heat from organic wastes, as in refuse-derived fuel.

Energy source. That from which energy can be derived.

Entropy. A measure of the randomness or disorder in a system.

Environmental resistance. Sum total of factors, such as predation and competition in the external environment, that limit the numerical increase of a population in a particular area.

Environmental science. The science that applies scientific facts and theories to solve definite environmental problems.

Equal protection laws. This Constitutional clause provides that equal protection and security shall be given to all under similar circumstances in one's life, liberty, and property. Therefore, the state cannot enact laws which are discriminatory.

Equilibrium. A stable condition in which net change does not occur, because opposing changes occur at exactly balancing rates. See dynamic equilibrium.

Essential amino acids. Amino acids that are not synthesized by the organism but are required in its diet.

Experiment. A formal test of a hypothesis under controlled or field conditions.

Exponential growth. Population growth in which the total number increases in the same manner as compounded interest, that is: 1, 2, 4, 8, 16, etc.

Extinction. The progressive deaths of all members of a particular species.

Fat. See lipids.

Fauna. The animals inhabiting a given area.

Feedback. See negative feedback and positive feedback.

First Law of Thermodynamics. Energy can be transformed to mass, and mass transformed to energy, but neither can be destroyed nor created.

Food chain. A linear sequence of food levels, each of which feeds on the previous one to acquire matter and energy.

Food pyramid. The feeding relationship between populations in a community; producers are at the base of the pyramid and the final consumer is at the top.

Food web. The complex feeding relationships among populations within a community.

Foreseeability. The ability to see or know in advance; hence, the reasonable anticipation that harm or injury is a likely result of acts or omissions.

Fossil fuels. The remains of once-living species which are burned to release energy; for example, coal, petroleum, and natural gas.

Free energy. Energy which is available to do work.

Gaia. This hypothesis postulates that the earth's entire ecosphere is analogous to a single living organism.

Glycogen. A polysaccharide constituting the principal form in which carbohydrate is stored in animals; animal starch.

Greenhouse effect. The heating of the atmosphere by virtue of the fact that short wavelength solar radiation is transmitted rather freely through the atmosphere and infrared radiation from the earth is more readily absorbed.

Growth rate. The change in population size over time.

Habitat. The place where a plant or animal normally lives or where individuals of population live.

Heat convection. The transfer of the vibrational energy of molecules, which constitutes heat energy, by the mechanism of molecular contact.

Heat energy. The kinetic energy associated with the random movement of molecules. The temperature of a substance depends on the average kinetic energy of component molecules. When heat is added to a substance, average kinetic energy increases.

Homeostasis. The capacity of a system to maintain a constant internal environment while external conditions vary.

Hydrocarbon. A compound of hydrogen and carbon that burns in the air to form water and oxides of carbon.

Hypervolume. The mathematical multidimensions of space. For example, a species may be described as a point in a space of many dimensions, one for each environmental factor involved. Such a mathematical space would be n-dimensional (more than three dimensions) or a hypervolume.

Hypothesis. In science, a statement or conclusion based on inductive reasoning about prior observations, which can be tested by an experiment.

Indirect cause. In law, an intervening force that came into motion after an act and combined with the act to cause injury.

Induction. Reasoning from the particular to the general. That is, deriving a general statement (hypothesis) based on individual observations.

Infrared radiation. Electromagnetic radiation lying outside the red band with wavelengths between approximately 0.8 to 1,000 micrometers.

Inorganic. Generally, not containing carbon in the form of rings or chains.

Instability. A property of any system that with certain disturbances or perturbations will increase in magnitude.

Intrinsic rate of increase. The rate at which a population of a particular species is capable of increasing under optimum conditions.

J-shaped growth curve. A population cycle characterized by rapid exponential growth followed by a sudden large crash in population size; characteristic of seasonal species such as small insects and some populations of small rodents like lemmings. Compare S-shaped growth curve.

Kwashiorkor. Malnutrition caused by diet high in carbohydrates and extremely low in protein.

Lipid. A group of organic compounds that are insoluble in water; notably fats, oils, and steroids.

Locucides. A factor that kills organisms indirectly by destroying their habitat or denaturing their locus.

Locus. The place or site at which a species is located in an environment.

Material conversion. Waste used in a different form, like getting road-paving material from auto tires.

Metabolism. The sum of all chemical and physical processes occurring within an organism.

Mortality. The rate at which members of a population die. In the case of human populations, the rate is normally given in deaths per thousand persons per annum.

Mutagen. An agent that is capable of increasing the rate of mutation.

Mutation. Change in genetic constitution of a species due to errors in the duplication of genes during meiosis. Errors can be induced by chemicals, radiation, or viruses.

Natality. The rate at which new individuals are produced in a population.

Natural resources. Energy and materials made available by geological processes.

Open system. A system in which constituents can move in and out of the system.

Optimum. Conditions which are most favorable.

Organic compound. A chemical compound containing carbon other than carbon dioxide, carbonic acid, and salts of carbonic acid. In general, organic compounds have molecules with carbon rings or chains.

Organism. An individual living thing.

Oxidation. A chemical reaction in which electrons are lost from an atom and its charge becomes more positive.

Ozone. An almost colorless (but faintly blue) gaseous form of oxygen made up of three oxygen atoms (O_3) and having an odor similar to weak chlorine. Ozone shields the earth's surface from ultraviolet radiation.

Pesticides. Chemical compounds used to control organisms, such as roaches and mosquitoes, which are deemed as pests.

Photochemical. A chemical reaction in which light is an agent or factor.

Plaintiff. One who commences a civil action.

Pollution. The condition caused by the addition to the water, air, or land of sufficient materials or waste heat so as to reduce or destroy their normal integrity. Pollution can also be caused by the overexploitation of natural resources.

Population. A group of organisms of the same species inhabiting a specified locality.

Positive feedback. A process whereby an increase in concentration of some output leads to an increase in its production. Unless ultimately controlled, positive feedback tends to be unstable.

Prima facie. A fact presumed to be true unless disproved by evidence to the contrary; presumably.

Probability. A mathematical expression of the degree of confidence that certain events will or will not occur.

Producers. In natural ecosystems, organisms like plants, able to synthesize their own food from inorganic substances. In anthroposystems, farms and industries would be characterized as producers.

Protein. Complex organic compounds built up from amino acids and constituting the major portion of organic matter in living protoplasm.

Proximate cause. That which in the ordinary course of events, unbroken by another or intervening cause, produces an injury and without which the injury would not have taken place.

Recycling. The reprocessing of wastes to recover the original raw material. For example, steel is recovered from tin cans and fiber is recovered from wastepaper.

Resource recovery. Refers to a productive use of material which would otherwise be disposed of as waste. It encompasses recycling, material conversion, and energy recovery.

Respiration. Chemical oxidative process whereby a living organism breaks down certain organic matter with the release of energy used in metabolism.

Reuse. Product is reused in the same form. For example, cleaning glass bottles.

Slash-and-burn agriculture. The practice of cutting down a small patch of forest, burning the plants, and cultivating the ash-enriched soil until its nutrients are depleted, at which time the area is abandoned in favor of another such region.

Smog. A term for air pollution, derived from the words "smoke" and "fog."

Solar constant. The amount of radiation from the sun falling upon a unit area per unit time at the distance of the earth: 1.95 cal./cm^2/min., or 1.36×10^6 erg/cm/sec.

S-shaped growth curve. Population growth curve which shows initial rapid expansion of the population when it is at low densities, then decelerating growth at higher densities, and an eventual leveling off as the density approaches the environment's carrying capacity (K). Compare J-shaped growth curve.

Stare decisis. The application by courts of rules announced in earlier decisions; let the decision stand.

Statute. Law enacted by a legislature.

Steady-state. An adjective describing a system that is in a stable dynamic state in which inputs balance outputs.

Substantive law. The branch of law that prescribes legal rights as opposed to the part which prescribes method of enforcing the rights or obtaining redress for their decisions.

Synergistic. The greater effect produced by two or more agents given simultaneously, than that would result from either alone.

System. Any interacting, interdependent, or associated group or objects in space and/or time.

Thermal pollution. Excessive increase in normal temperatures of natural waters caused by waste heat from industrial activities.

Thermodynamics. A branch of physics dealing with the study of heat and heat transfer.

Threatened species. A species likely to become endangered in the foreseeable future.

Toxic substances. A chemical or mixture of chemicals that is poisonous to organisms.

Trophic levels. The various feeding levels in a food chain, including producers, consumers, and decomposers.

Ultraviolet. Being in the portion of the electromagnetic spectrum that has frequencies somewhat higher than the violet end of the visible spectrum.

Variable. Any of the characteristics which vary or change with time, for example, species diversity, temperature, relative humidity, light.

Vitamin. An organic substance present in minute amounts in food and necessary in minute quantities for certain metabolic processes.

Waste-water treatment plant. A series of tanks, screens, filters, and other processes by which pollutants are removed from water.

Bibliography

Anderson, S. H., R. E. Beiswenger, and P. W. Purdom. 1987. *Environmental Science.* Columbus, Ohio: Merrill Publishing Company.

Ayres, R. V., and A. V. Kneese. 1971. Economic and ecological effects of a stationary economy. *Annual Review of Ecology and Systematics* 2:-22.

Bacastrow, R. B., and C. D. Keeling. 1981. Atmospheric carbon dioxide concentration, the observed airborne fraction, the fossil fuel airborne fraction, and the difference in hemispheric airborne fractions. In *Scope* 16: Global Carbon Modeling, B. Bolin (ed.). London: John Wiley and Sons.

Bacow, L. S., and M. Wheeler. 1984. *Environmental Dispute Resolution.* New York: Plenum Press.

Barney, G. O. 1979. *The Global 2000 Report to the President of the United States.* New York: Pergamon Press.

Begon, M., J. L. Harper, and C. R. Townsend. 1986. *Ecology: Individuals, Populations and Communities.* Sunderland, Mass.: Sinauer Associates, Inc.

Bennett, R. J., and R. J. Chorley. 1978. *Environmental Systems.* Princeton, N.J.: Princeton University Press.

Berelson, B. 1969. Beyond family planning. *Science* 163 (3867):533-43.

Beres, L. R. 1975. *Planning Alternative World Futures,* L. R. Beres and H. R. Targ (eds.). New York: Praeger.

Bergstrom, G. 1973. *The Food and People Dilemma.* North Scituate, Mass.: Duxbury Press.

Bilder, R. B. 1980. International law and natural resources policies. *Natural Resources Journal* 20:451.

Black, B., and D. E. Lilienfeld. 1984. Epidemiologic proof in toxic tort litigation. *Fordham Law Review* 52 (April):732-85.

Black's Law Dictionary. 1979. St. Paul, Minn.: West Publishing Company.

Boulding, K. E. 1966. *The Economics of the Coming Spaceship Earth.* Resources for the Future Book. Baltimore: Johns Hopkins Press.

———. 1982. Knowledge, resources and the future. *BioScience* 32:343-44.

Bowman, K. P. 1988. Global trends in total ozone. *Science* 239:48-50.

Brewer, R. 1988. *The Science of Ecology.* New York: W. B. Saunders Company.

Brown, L. R. 1971. Quoted in "Is There an Optimum Level of Population?" S. F. Singer (ed.). Hightstown, N.J.: McGraw-Hill.

———. 1976. World population trends: Signs of hope, signs of stress. *Worldwatch Paper* 8. Washington, D.C.: Worldwatch Institute.

———. 1981. *Building a Sustainable Society.* New York: W. W. Norton.

———. 1982. Living and working in a sustainable society. *Futurist* 16, 66.

———. 1984. Putting food on the world's table: A crisis of many dimensions. *Environment* 26(4):15-20, 38-43.

———. 1987. Analyzing the demographic trap. *The State of the World 1987,* L. R. Brown et al. (eds.). New York: W. W. Norton.

———. 1988. *Breakthrough on Soil Erosion.* Washington, D.C.: Worldwatch Institute.

———. 1989. Reexamining the world food prospect. *The State of the World 1989,* L. R. Brown et al. (eds.). New York: W. W. Norton.

Brown, L. R., and E. C. Wolf. 1987. Charting a sustainable course. *The State of the World 1987,* L. R. Brown et al. (eds.). New York: W. W. Norton.

Brown, L. R., C. Flavin, and S. Postel. 1989. Outlining a global plan. *The State of the World 1989,* L. R. Brown et al. (eds.). New York: W. W. Norton.

Bugliarelli, G. 1978. A technological magistrature. *Bulletin of the Atomic Scientist* 34(1):34-37.

Cairns, J., Jr. 1978. Quantification of biological integrity. *The Integrity of Water,* R. K. Ballentine and L. J. Guarria (eds.), U.S.E.P.A., Office of Water and Hazardous Materials, Washington, D.C.: Superintendent of Documents.

Caldwell, L. K. 1984. *International Environmental Policy: Emergence and Directions.* Durham, N.C.: Duke University Press.

Campbell, A. B. 1977. *Physics Today,* Dec., p. 38.

Catton, W. R., Jr. 1987. The world's most polymorphic species. *BioScience* 37(6):413-19.

Chapman, R. N. 1928. The quantitative analysis of environmental factors. *Ecology* 9:111-22.

Churchman, C. W. 1971. *The Design of Inquiring Systems: Basic Concepts of Systems and Organizations.* New York: Basic Books.

Clapham, W. B., Jr. 1981. *Human Ecosystems.* N.Y.: Macmillan Publishing Company.

Clark, C. W. 1981. Bionomics. *Theoretical Ecology* (2nd ed.), R. M. May (ed.). Sunderland, Mass.: Sinauer Associates.

Cloud, P. E., Jr. 1969. Quoted in: *Effects of Population Growth on Natural Resources and the Environment.* Hearings before a Subcommittee of the Committee on Government Operations, House of Representatives. Washington, D.C.: Superintendent of Documents.

Coale, A. J. 1970. Man and his environment. *Science* 170:132-36.

―――. 1974. The history of the human population. *Scientific American* 231(3):41-51.

Cohn, J. P. 1987. Chlorofluorocarbons and the ozone layer. *BioScience* 37:647-50.

Cole, L. C. 1958. The ecosphere. *Scientific American* 198(4):83-92.

Commission on Population Growth. 1972. *Population and the American Future.* Washington, D.C.: Superintendent of Documents.

Committee on Government Operations. 1969. *Effects of Population Growth on Natural Resources and the Environment.* Washington, D.C.: Superintendent of Documents.

Conable, B. B. 1988. Address to the Board of Governors, Berlin, September 27, 1988. Cited in Brown et al. 1989.

Conservation Foundation. 1983. *Public Policy, Science, and Environmental Risk.* Washington, D.C.: The Brookings Institution.

Contini, P., and P. H. Sand. 1972. Methods to expedite environmental protection: International ecostandards. *American Journal of International Law* 66:35.

Cooper, A. W. 1982. Why doesn't anyone listen to ecologists―and what can ESA do about it? *Bulletin of Ecological Society of America* 63(4):348, 350.

Couloumbis, T. A., and J. H. Wolfe. 1982. *International Relations: Power and Justice.* Englewood Cliffs, N.J.: Prentice-Hall.

Council on Environmental Quality. 1979. *Environmental Quality: The Tenth Annual Report.* Washington, D.C.: Superintendent of Documents.

―――. 1981. *Environmental Trends.* Washington, D.C.: Superintendent of Documents.

―――. 1982. *The Global 2000 Report.* Washington, D.C.: Superintendent of Documents.

————. 1984. *Environmental Quality: The Fifteenth Annual Report.* Washington, D.C.: Superintendent of Documents.

————. 1985. *Environmental Quality: The Sixteenth Annual Report.* Washington, D.C.: Superintendent of Documents.

Daly, H. E. 1973. *Toward a Steady-State Economy.* San Francisco: Freeman & Co.

————. 1974. The economics of the steady state. *American Economic Review* 64:15.

————. 1986. Toward a new economic model. *Bulletin of the Atomic Scientists* 42:42-44.

Deevey, E. S. 1956. *Scientific American* 194:105.

————. 1960. The human population. *Scientific American* 203(3), Offprint No. 608. San Francisco: W. H. Freeman & Co.

Doninger, D. D. 1978. Federal regulation of vinyl chloride: A short course in the law and policy of toxic substances. *Ecology Law Quarterly* 7:500-21.

Ehrenfeld, D. W. 1976. The conservation of non-resources. *American Scientist* 64:648-56.

Ehrlich, P. R., and J. P. Holdren. 1971. Impact of population growth. *Science* 171:1212-16.

Enger, E. D., J. R. Kormelink, B. F. Smith, and R. J. Smith. 1989. *Environmental Science: The Study of Interrelationships.* Dubuque, Iowa: William C. Brown Publishers.

Falk, R. 1971. *This Endangered Planet.* New York: Random House.

Farman, J. C., B. G. Gardiner, and J. D. Shanklin. 1985. Large ozone losses in Antarctica reveal seasonal CC10x/NOx interaction. *Nature* 415:207-10.

Feld, W. J., and R. S. Jordan. 1983. *International Organizations: A Comparative Approach.* New York: Praeger.

Fitzgerald, S. G. 1986. World Bank pledges to protect wildlands. *BioScience* 36:712-15.

Forrester, J. W. 1973. Counterintuitive behavior of social systems. *Toward Global Equilibrium: Collected Papers,* D. L. Meadows and D. H. Meadows (eds.). Cambridge, Mass.: Wright-Allen Press.

————. 1982. Global modeling revisited. *Futures* 14:95-110.

Freese, L. 1985. Social traps and dilemmas: Where social psychology meets human ecology. American Sociological Association Meeting at Washington State University.

Friedmann, W. G. 1972. *Law in a Changing Society.* New York: Columbia University Press.

Frisch, R. E. 1978. Population, food intake and fertility. *Science* 199:22-30.

Galloway, J. N., Z. Dianwu, J. Xiong, and G. E. Likens. 1987. Acid rain: China, United States, and a remote area. *Science* 236:1559-62.

Georgescu-Roegen, N. 1977. The steady-state and ecological salvation: A thermodynamic analysis. *BioScience* 27:266-71.

Gerlach, L. P., and V. H. Hine. 1973. *Lifeway Leap*. Minneapolis: University of Minnesota Press.

Gold, S. 1986. Causation in toxic torts: Burdens of proof, standards of persuasion, and statistical evidence. *Yale Law Journal* 96 (December): 376-402.

Gorse, J. E., and D. R. Steeds. 1987. *Desertification in the Sahelian and Sudanian Zones of West Africa*. Washington, D.C.: World Bank.

Goudie, A. 1986. *The Human Impact on the Natural Environment*. Cambridge, Mass.: MIT Press.

Grieves, F. L. 1975. *Environmental Affairs* 4:309.

_____. 1977. *Conflict and Order*. Boston: Houghton Mifflin Co.

Grunchy, A. G. 1947. *Modern Economic Thought: The American Contribution*. Englewood Cliffs, N.J.: Prentice-Hall.

Gumbel, E. J. 1941. Probability-interpretation of the observed return periods of floods. *Transactions of the American Geophysical Union* 22:836.

Hammond, N. 1986. The emergence of Mayan civilization. *Scientific American* 255(2):106-15.

Hanks, J. 1988. Southern Africa's abused environment. *Earthwatch*, No. 31.

Hardesty, D. L. 1977. *Ecological Anthropology*. New York: John Wiley & Sons.

Hardin, G. 1968. Tragedy of the commons. *Science* 162:1243-48.

Harlan, J. R. 1976. The plants and animals that nourish man. *Scientific American* 235(3):88-97.

Harwell, M. A., and C. C. Harwell. 1989. Environmental decision making in the presence of uncertainty. *Ecotoxicology: Problems and Approaches*, S. A. Levin et al. (eds.). New York: Springer-Verlag.

Heck, W. W., et al. 1983. A reassessment of crop loss from ozone. *Environmental Science and Technology* 17(12):573A-81A.

Henderson-Sellers, A., and K. McGuffie. 1986. The threat from melting icecaps. *New Scientist* (June 12).

Hill, A. R. 1975. Ecosystem stability in relation to stress caused by human activity. *Canadian Geographer* 19:206.

Hill, J., and S. L. Durham. 1978. Input, signals and control in ecosystems. Proceedings 1978 Conference on Acoustics, Speech and Signal Processing, Tulsa, Okla. New York Institute of Electrical and Electronics Engineers, 391-97.

Hill, J., and R. G. Wiegert. 1980. Microcosms in ecological modeling. *Microcosms in Ecological Research*, J. P. Giesy (ed.). U.S. Department of Energy Symposia 52. Springfield, Va.: National Technical Information Service, pp. 138-63.

Hutchinson, G. E. 1958. Concluding remarks. *Cold Spring Harbor Symposia on Quantitative Biology* 22:415.

Jacobsen, T., and R. M. Adams. 1958. Salt and silt in ancient Mesopotamian agriculture. *Science* (November 21).

Jacobson, J. L. 1987. Planning the Global Family. *Worldwatch Paper* 80. Washington, D.C.: Worldwatch Institute.

Josephson, J. 1982. Why maintain biological diversity? *Environmental Science and Technology* 16:94A-97A.

Kaplan, M. A., and N. B. Katzenbach. 1961. *The Political Foundation of International Law.* New York: Wiley.

Kieffer, G. H. 1979. *Bioethics, A Textbook of Values.* Reading, Mass.: Addison-Wesley.

Kile, quoted in Forrester (1982).

Kuhn, A. 1974. *The Logic of Social Systems.* San Francisco: Jossey-Bass.

Kuhn, T. 1962. *The Structure of Scientific Revolutions.* Chicago: University of Chicago Press.

Kupchella, C. E., and M. C. Hyland. 1989. *Environmental Science: Living Within the System of Nature.* Needham Heights, Mass.: Allyn and Bacon.

Langford, T. E. 1972. A Comparative Assessment of Thermal Effects in Some British and North American Rivers. In R. T. Oglesby, C. A. Carlson and J. A. McCann (eds.), *River Ecology and Man.* New York: Academic Press, 318-51.

Lastrucci, C. L. 1967. *The Scientific Approach—Basic Principles of the Scientific Method.* Cambridge, Mass.: Schenkman.

Leopold, A. 1949. *The Land Ethic.* New York: Oxford University Press.

Lovelock, J. E. 1979. *Gaia: A New Look at Life on Earth.* New York: Oxford University Press.

Lowe, J.W.G. 1985. *The Dynamics of Apocalypse.* Albuquerque, N.M.: University of New Mexico Press.

Luoma, S. N. 1984. *Introduction to Environmental Issues.* New York: Macmillan Publishing Co.

Malthus, T. R. 1816. *Essay on the Principle of Population.* Totowa, N.J.: Biblio Distribution Center.

Manabe, S., and R. T. Wetherald. 1986. Reduction in summer soil wetness induced by an increase in atmospheric carbon dioxide. *Science* (May 2).

Maugh, T. H. 1980. Ozone depletion would have dire effects. *Science* 207:394-95.

Mauldin, W. P. 1980. Population trends and prospects. *Science* 209:143-57.

McElroy, M. B., et al. 1986. Reductions of Antarctic ozone due to synergistic interactions of chlorine and bromine. *Nature* 321:759-62.

McHale, J., and M. C. McHale. 1976. *Human Requirements, Supply Levels and Outer Bounds: A Framework for Thinking about the Planetary Bargain.* Palo Alto, Calif.: Aspen Institute for Humanistic Studies.

Meadows, D. H., et al. 1972. *Limits to Growth: A Global Challenge.* New York: Universe Books.

Measarovic, M., and E. Pestel. 1974. *Mankind at the Turning Point.* London: Hutchinson.

Miller, G. T., Jr. 1986. *Environmental Science: An Introduction.* Belmont, Calif.: Wadsworth Publishing Co.

Mitchell, M., Jr. 1978. *Prospect for Man: Climate Change,* J. R. Miller, ed. Toronto: York University.

Mott, R. 1988. An acid rain summons from Europe. *The Environmental Forum,* March/April.

Myers, N. 1980. The problem of disappearing species: What can be done? *Ambio* 19:229-35.

Nanda, V. P., and P. T. Moore. 1983. Global management of the environment: Regional and multilateral initiatives. In: *World Climate Change,* V. P. Nanda (ed). Boulder, Colo.: Westview Press.

National Academy of Sciences. 1969. *Resources and Man.* San Francisco: W. H. Freeman.

_____. 1979. *Carbon Dioxide and Climate: A Scientific Assessment.* Washington, D.C.: National Academy of Sciences.

National Research Center for Toxicological Research. 1979. The Food and Drug Administration—The ED01 Study. *Journal of Experimental Pathology and Toxicology.*

National Research Council. 1977. Pesticide Information Review and Evaluation Committee/Assembly of Life Sciences. An Evaluation of the Carcinogenicity of Chlordane and Heptachlor. Washington, D.C.: National Academy of Sciences.

_____. 1980. Recommended dietary allowances. 9th revised edition. Washington, D.C.: National Academy Press.

Nebel, B. J. 1987. *Environmental Science: The Way the World Works.* Englewood Cliffs, N.J.: Prentice-Hall.

Nelson, H., and R. Jurmain. 1985. *Introduction to Physical Anthropology.* St. Paul, Minn.: West.

Notestein, F., et al. 1963. The problem of population control. *The Population Dilemma,* P. M. Hauser (ed.) Englewood Cliffs, N.J.: Prentice-Hall.

Occupational Safety and Health Administration. 1979. Identification,

Classification and Regulation of Toxic Substances Posing a Potential Occupational Carcinogenic Risk. *Federal Register* 42.

Ochero, L. 1981. Newsletter of October 1981. Population Reference Bureau.

Odum, E. P. 1971. *Fundamentals of Ecology.* Philadelphia: Saunders.

———. 1983. *Basic Ecology.* Philadelphia: Saunders.

———. 1989. *Ecology and Our Endangered Life-Support Systems.* Sunderland, Mass.: Sinauer Associates, Inc.

Oliver, B. G., and A. J. Niimi. 1985. Bioconcentration factors of some halogenated organics for rainbow trout: Limitations in their use for prediction of environmental residues. *Environmental Science and Technology* 19:842-49.

Olmstead v. United States. 1928. 227 U.S. 438, 479.

Ophuls, W. 1977. *Ecology and the Politics of Scarcity.* San Francisco: W. H. Freeman.

Ottar, B. 1977. International agreement needed to reduce long-range transport of air pollutants in Europe. *Ambio* 6(5):262-69.

Paarlberg, D. 1988. *Toward a Well-Fed World.* Ames, Iowa: Iowa State University Press.

Patrick, R., et al. 1981. Acid lakes from natural and anthropogenic causes. *Science* 211:446-48.

Pattee, H. H. 1973. *Hierarchy Theory.* H. H. Pattee (ed.). New York: Braziller.

Peccei, A. 1982. Global modeling for humanity. *Futures* 14:91-94.

Pimentel, D., et al. 1987. World agriculture and soil erosions. *BioScience* (April).

Population Reference Bureau. 1985. World Population Data Sheet. Washton, D.C.: Population Reference Bureau, Inc.

———. 1988. World Population Data Sheet. Washington, D.C.: Population Reference Bureau, Inc.

Postel, S. 1984. *Air Pollution, Acid Rain and the Future of Forests.* Worldwatch Institute Paper 58. Washington, D.C.: Worldwatch Institute.

———. 1989. Halting land degradation. In: *State of the World,* Brown, L. R., et al. (eds.). New York: W. W. Norton.

Potter, V. R. 1971. *Bioethics, Bridge to the Future.* Englewood Cliffs, N.J.: Prentice-Hall.

———. 1988. *Global Bioethics: Building on Leopold Legacy.* East Lansing, Mich.: Michigan State University Press.

Pratt, W. E. 1952. Toward a philosophy of oil-finding. *Bulletin of American Association of Petroleum Geologists* 36:2231-36.

President's Commission on Population Growth and the American Future. 1972. Washington, D.C.: Superintendent of Documents.

Puchala, D. J. 1988. The United Nations and ecosystem issues: Institutionalizing the global interest. *Politics in the United Nations System,* Finkelstein, L. S. (ed.). Durham, N.C.: Duke University Press.

Rahn, K. A., and D. Lowenthal. 1984. Elemental tracers of distant regional pollution aerosols. *Science* 223:132-39.

Raup, D. M. 1986. Biological extinction in earth history. *Science* 231:1528-33.

ReVelle, P., and C. ReVelle. 1984. *The Environment: Issues and Choices for Society.* Boston: Willard Grant Press.

Rolston, H. 1986. *Philosophy Gone Wild.* Buffalo, N.Y.: Prometheus Books.

Rosen, R. 1975. Biological systems as paradigms for adaptation. *Adaptive Economic Models,* Day, R. H. and Groves, T. (eds.). New York: Academic Press.

Rosencranz, A. 1980. The problem of transboundary pollution. *Environment* 22:5, 15-20.

_____. 1983. The international law and politics of acid rain. *World Climate Change,* Nanda, V. P. (ed.). Boulder, Colo.: Westview Press.

Ruckelshaus, W. 1983. Risk Assessment in the Federal Government. Washington, D.C.: National Academy of Sciences Report.

Ruggieri, G. D. 1976. Drugs from the sea. *Science* 194:491-97.

Sagoff, M. 1985. Fact and value in ecological science. *Environmental Ethics* 7:99-116.

Santos, M. A. 1974. Ecological systems versus human systems: Which should be supreme? *Journal of Environmental Systems* 4:261-67.

_____. 1983. Quantification of the anthroposystem concept. *Journal of Environmental Systems* 12:351-61.

Schindler, D. W., et al. 1985. Long-term ecosystem stress: The effects of years of experimental acidification on a small lake. *Science* 228:1395-96.

Schleidegger, A. E. 1975. *Physical Aspects of Natural Catastrophes.* New York: Elsevier.

Schneider, S. H. 1983. Food and climate: Basic issues and some policy implications. *World Climate Change,* Nanda, V. P. (ed.). Boulder, Colo.: Westview Press.

Schware, R., and W. W. Kellogg. 1983. International strategies and institutions for coping with climate change. *World Climate Change,* Nanda, V. P. (ed.). Boulder, Colo.: Westview Press.

Sears, P. B. 1957. Man the newcomer: the living landscape and a new tenant. *Man's Natural Environment, a System Approach,* L. H. and E. Sommerville (eds.). North Scituate, Mass.: Duxbury.

Shea, C. P. 1989. Protecting the ozone layer. *State of the World 1989,* L. R. Brown (ed.). New York: W. W. Norton.

Shelby, B., and T. Heberlein. 1984. A conceptual framework for carrying capacity determination. *Leisure Sciences* 6:433-51.

Shevardnadze, E. A. 1988. Statement before the Forty-third Session of the U.N. General Assembly (September 27). New York.

Shevchenko, A. 1985. *Breaking with Moscow.* New York: Alfred A. Knopf.

Simon, J. L. 1984. Bright global future. *The Bulletin of the Atomic Scientists* 40:14-17.

Simon, J. L., and H. Kahn. 1984. *The Resourceful Earth: A Response to Global 2000 Report.* New York: Basil Blackwell.

Singer, S. F. 1971. *Is There an Optimum Level of Population,* S. F. Singer (ed.). Hightstown, N.J.: McGraw-Hill.

Skarby, L., and G. Sellden. 1984. The effects of ozone on crops and forests. *Ambio* 13:68-72.

Sontag, J. M. 1977. Aspects in carcinogen bioassay. *Origins of Human Cancer,* H. H. Hiatt, et al. (eds.). Cold Spring Harbor, N.Y.: Cold Spring Harbor Laboratory.

Southwick, C. H. 1985. Environmental impacts of early societies and the rise of agriculture. *Global Ecology,* C. H. Southwick (ed.). Sunderland, Mass.: Sinauer Associates, Inc.

Spengler, J. J. 1971. *Is There an Optimum Level of Population?* S. F. Singer (ed.). Hightstown, N.J.: McGraw-Hill.

Spitler, A. 1987. Exchanging debt for conservation. *BioScience* 37:781.

Sprout, H., and M. Sprout. 1971. *Toward a Politics of the Planet Earth.* New York: Van Nostrand Reinhold.

Steel, R.G.D., and J. H. Torrie. 1960. *Principles and Procedures of Statistics.* New York: McGraw-Hill.

Stone, C. D. 1974. *Should Trees Have Standing?* Los Altos, Calif.: William Kaufmann, Inc.

Stone, J. 1984. *Visions of World Order.* Johns Hopkins University Press.

Strong, M. 1987. Beyond Foreign Aid—Towards a New World System. Presented to the International Development Conference (March 19). Washington, D.C.

Surgeon General. 1988. The Surgeon General's Report on Nutrition and Health. Washington, D.C.: Superintendent of Documents.

Tansley, A. G. 1935. The use and abuse of vegetational concepts and terms. *Ecology* 16:284-307.

Teitelbaum, M. S. 1975. Relevance of demographic transition theory for developing countries. *Science* 188:420-26.

Tomatis, L., et al. 1973. The predictive value of mouse liver tumor induction in carcinogenicity testing—a literature survey. *International Journal of Cancer* 12:1-20.

Trabalka, J. R. 1985. Atmospheric Carbon Dioxide and the Global Carbon Cycle. U.S. Department of Energy, DOE/Er-0239.

Trail Smelter Arbitration (*United States v. Canada*), 3 R. International Arbitration Awards 1911 (1938); 33 *American Journal of International Law* 182 (1939); 35 *American Journal of International Law* 684 (1941).

Train, R. 1978. The environment today. *Science* 201:320-24.

Turk, J. 1989. *Introduction to Environmental Studies.* New York: Saunders.

United Nations. 1967. *United Nations Universal Declaration of Human Rights.* U.N. Publications.

_____. 1969. Charter of the United Nations, as amended 31 December 1969.

_____. 1988. Statistical Office, Monthly Bulletin of Statistics, New York, September.

United Nations Environmental Programme. 1984. *General Assessment of Progress in the Implementation of the Plan of Action to Combat Desertification.* New York: United Nations.

United Nations Food and Agriculture Organization. 1984. New York: United Nations.

United States Bureau of Census. 1985. United States Department of Commerce.

_____. 1987. United States Department of Commerce.

United States Department of Agriculture. 1988. *Cropland, Water, and Conservation Situation and Outlook Report* (September). Economic Research Service. Washington, D.C.

Vago, S. 1981. *Law and Society.* Englewood Cliffs, N.J.: Prentice-Hall.

Valaskakis, K. 1981. The conserver society: Emerging paradigm of the 1980's? *The Futurist* 15(5):5-13.

Vallis, L. B., and M. Naturajan. 1987. The stratosphere 1979-1984: Large infusions of odd nitrogen during solar cycle 21. *Nature* 28.

Vattel 1835. *The Law of Nations 6* (4th American Edition). J. Chiity (ed.).

Verstraete, M. M. 1986. Defining desertification: A review. *Climate Change,* No. 9.

Wagar, J. A. 1964. The carrying capacity of wild lands for recreation. *Forestry Science Monographs* 7:1-24.

Watson, R. T., et al. 1986. Present state of knowledge of the upper atmosphere: An assessment report. NASA Reference Publication 1162.

Watt, K.E.F. 1968. *Ecology and Resource Management.* New York: McGraw-Hill.

_____. 1982. *Understanding the Environment.* Boston: Allyn and Bacon.

Weinberg, A. M., and R. P. Hammond. 1972. *Bulletin of the Atomic Scientists* 43-44:5.

Westhoff, C. F. 1978. Marriage and fertility in the developed countries. *Scientific American* 239:51-57.

Westing, A. H. 1981. A world in balance. *Environmental Conservation* 8:177.

Weston, B. H., et al. 1980. *International Law and World Order.* St. Paul, Minn.: West Publishing Co.

Whitney, S. 1973. The case for creating a special environmental court system—further comment. *William and Mary Law Review* 15:33.

Wilson, E. O., and W. H. Bossert. 1971. *A Primer of Population Biology.* Sunderland, Mass.: Sinauer Associates, Inc.

Winikoff, B. 1978. Nutrition, population and health: Some implications for policy. *Science* 200:895-902.

Wolf, E. C. 1988. Avoiding a mass extinction of species. *State of the World 1988,* R. L. Brown, et al. (eds.). New York: W. W. Norton.

Woodwell, G. M. 1967. Toxic substances and ecological cycles. *Scientific American* 216(3):24-31.

———. 1978a. The carbon dioxide question. *Scientific American* 238:1, 34-43.

———. 1978b. The nature of the deforestation problem—trends and policy implication. Proceedings of the U.S. Strategy Conference on Tropical Deforestation, U.S. Dept. of State and U.S. Agency for International Development, Washington, D.C.

World Bank. 1984. World Development Report. Washington, D.C.

———. 1985. Desertification in the Sahelian and Sudanian Zones of West Africa. Washington, D.C.

World Health Organization. 1974. Assessment of the carcinogenicity and mutagenicity of chemicals. Technical Report Series, No. 546, pp. 9-11.

Worldwatch Institute, 1978. The Global Environment and Basic Human Needs: A Report to the Council on Environmental Quality. Washington, D.C.: Worldwatch Institute.

Ziman, J. 1984. *An Introduction to Science Studies.* Cambridge, England: Cambridge University Press.

Zimmerman, F. 1984. Summary Record of the Multilateral Conference on the Environment, FRG government, Munich.

Name Index

Subject Index

ABOUT THE AUTHOR

MIGUEL A. SANTOS is an Associate Professor of Ecology and Biology at Baruch College of the City University of New York. As the coordinator of the Environmental Studies Program he has developed a series of courses and seminars on environmental issues. He received his Ph.D. in Ecology and Zoology from Rutgers–The State University of New Jersey and J.D. from the Law School of Rutgers University. Dr. Santos is an interdisciplinary scholar, with research interests in biology, ecology, environmental sciences, and law. In addition to publishing numerous articles in scholarly journals, he is the author of *Ecology, Natural Resources, and Pollution; Genetics and Man's Future: Legal, Social, and Moral Implications of Genetic Engineering; Laboratory Manual for Environmental Studies;* and *Ecology, Environment, and Society.* Dr. Santos has served as a scientific-legal consultant for both research and industrial organizations.